ECONOMICS OF POLICY OPTIONS TO ADDRESS CLIMATE CHANGE

ECONOMICS OF POLICY OPTIONS TO ADDRESS CLIMATE CHANGE

GREGORY N. BARTOS

Nova Science Publishers, Inc.
New York

Copyright © 2009 by Nova Science Publishers, Inc.

All rights reserved. No part of this book may be reproduced, stored in a retrieval system or transmitted in any form or by any means: electronic, electrostatic, magnetic, tape, mechanical photocopying, recording or otherwise without the written permission of the Publisher.

For permission to use material from this book please contact us:
Telephone 631-231-7269; Fax 631-231-8175
Web Site: http://www.novapublishers.com

NOTICE TO THE READER
The Publisher has taken reasonable care in the preparation of this book, but makes no expressed or implied warranty of any kind and assumes no responsibility for any errors or omissions. No liability is assumed for incidental or consequential damages in connection with or arising out of information contained in this book. The Publisher shall not be liable for any special, consequential, or exemplary damages resulting, in whole or in part, from the readers' use of, or reliance upon, this material.

Independent verification should be sought for any data, advice or recommendations contained in this book. In addition, no responsibility is assumed by the publisher for any injury and/or damage to persons or property arising from any methods, products, instructions, ideas or otherwise contained in this publication.

This publication is designed to provide accurate and authoritative information with regard to the subject matter covered herein. It is sold with the clear understanding that the Publisher is not engaged in rendering legal or any other professional services. If legal or any other expert assistance is required, the services of a competent person should be sought. FROM A DECLARATION OF PARTICIPANTS JOINTLY ADOPTED BY A COMMITTEE OF THE AMERICAN BAR ASSOCIATION AND A COMMITTEE OF PUBLISHERS.

LIBRARY OF CONGRESS CATALOGING-IN-PUBLICATION DATA

Available upion request.

ISBN: 978-1-60692-116-6

Published by Nova Science Publishers, Inc. ✢ New York

CONTENTS

Preface		**vii**
Abbreviations		**ix**
Correspondence		**xi**
Chapter 1	**Results in Brief**	**1**
Chapter 2	**Background**	**5**
	Mitigating Greenhouse Gas Emissions	11
	Adaptation Policies to Reduce Vulnerability to Climate Change	13
	Estimating the Potential Benefits and Costs of Actions to Address Climate Change	14
	Climate Change Policies Currently under Consideration by the Congress	16
Chapter 3	**Despite some Uncertainty Regarding the Potential Economic Impact, All of the Panelists Supported Establishing a Price on Greenhouse Gas Emissions**	**17**
	All of the Panelists Agreed that the Congress Should Consider Establishing a Price on Greenhouse Gas Emissions, and the Majority Recommended Complementary Policies	18
	Panelists Described the Potential Benefits and Costs of Actions to Address Climate Change	22

	Panelists Rated Estimates of Costs of Actions to Address Climate Change as more Useful than Estimates of Benefits, Citing Uncertainties Associated with Future Impacts	25
	The Majority of Panelists Said that the United States should Begin to Control Emissions Soon, regardless of International Participation	27
Chapter 4	**The Panelists' Views on the Strengths and Limitations of Policy Options Focused Primarily on the Environmental Certainty of a Cap-and-Trade System versus the Efficiency of a Tax on Emissions**	29
	Panelists Rated the Importance of Criteria for Evaluating Policy Options	31
	The Most Important Trade-offs Are between the Relative Effectiveness of Cap-and-Trade Systems and the Relative Efficiency of Taxes	31
	The Panelists Viewed Other Policy Options less Favorably but Cited Their Potential as a Complement to a Market-Based Mechanism	33
Appendix I	**Scope and Methodology**	39
Appendix II	**Selected Characteristics of Panelists' Preferred Policy Options for Addressing Climate Change**	43
Appendix III	**Selected Questions and Expert Responses**	47
Appendix IV	**Panel of Experts**	57
Appendix V	**Bibliography of Selected Literature Reviewed by GAO**	59
Endnotes		71
Index		75

PREFACE

Elevated levels of greenhouse gases in the atmosphere and the resulting effects on the earth's climate could have significant environmental and economic impacts in the United States and internationally. Potential impacts include rising sea levels and a shift in the intensity and frequency of floods and storms. Proposed responses to climate change include adapting to the possible impacts by planning and improving protective infrastructure, and reducing greenhouse gas emissions directly through regulation or the promotion of low-emissions technologies. Because most U.S. emissions stem from the combustion of fossil fuels such as coal, oil, and natural gas, much of this book centers on the effect emissions regulation could have on the economy.

ABBREVIATIONS

CCSP	Climate Change Science Program
CCTP	Climate Change Technology Program
DOE	Department of Energy
EPA	Environmental Protection Agency
EU ETS	European Union Emissions Trading Scheme
IPCC	Intergovernmental Panel on Climate Change
NAS	National Academy of Sciences
NASA	National Aeronautics and Space Administration
NOAA	National Oceanic and Atmospheric Agency
UNFCCC	United Nations Framework Convention on Climate Change
USDA	U. S. Department of Agriculture

CORRESPONDENCE

May 9, 2008

The Honorable Barbara Boxer
Chairman
Committee on Environment and Public Works
United States Senate

The Honorable Dianne Feinstein
United States Senate

Changes in the earth's climate attributable to increased concentrations of greenhouse gases may have significant environmental and economic impacts in the United States and internationally.[1] Among other potential impacts, climate change could threaten coastal areas with rising sea levels, alter agricultural productivity, and increase the intensity and frequency of floods and tropical storms. Furthermore, climate change has implications for the fiscal health of the federal government, affecting federal crop and flood insurance programs, and placing new stresses on infrastructure and natural resources.[2]

The earth's climate system is driven by energy from the sun and is maintained by complex interactions among the atmosphere, the oceans, and the reflectivity of the earth's surface, among other factors. Certain gases in the earth's atmosphere — such as carbon dioxide and methane — are known as greenhouse gases because they trap energy from the sun and prevent it from returning to space. Climate change is a long-term and global issue because greenhouse gases disperse widely in the atmosphere once emitted and can remain for extended periods of time. According to the Intergovernmental Panel on Climate Change (IPCC) — an organization within the United Nations that assesses scientific, technical, and economic information on the effects of climate change — atmospheric concentrations of carbon dioxide, the most common greenhouse gas, rose 35 percent between pre-industrial times and 2005.[3] The IPCC has determined that 11 of the 12 warmest years on record occurred between 1995 and 2006 and expects that global mean temperatures will continue to rise over the next century as a result of increased atmospheric concentrations of greenhouse gases.

In 2006, carbon dioxide released from the burning of fossil fuels accounted for approximately 78 percent of anthropogenic greenhouse gas emissions in the United States. The remaining 22 percent of emissions included carbon dioxide from nonenergy use of fossil fuels and iron and steel production; methane from landfills, coal mines, oil and gas operations, and agriculture; nitrous oxide from fossil fuels, fertilizers, and industrial processes; and other gases emitted from processes such as refrigeration, air conditioning, and semiconductor manufacturing.[4] In 2005, the United States was the largest global emitter of carbon dioxide followed by China, Russia, Japan, and India.[5]

The United States, 190 other nations, and the European Economic Community have ratified the United Nations Framework Convention on Climate Change (the Framework Convention), which aims to stabilize atmospheric greenhouse gas concentrations within a time frame sufficient to allow ecosystems to adapt naturally to climate change, to ensure that food production is not threatened, and to enable economic development to proceed in a sustainable manner. Under the Kyoto Protocol of the Framework Convention, 178 nations have agreed to reduce their greenhouse gas emissions by at least 5 percent below 1990 levels, by 2012. Also, in 2005, the European Union began implementing its Emissions Trading Scheme (EU ETS), a program that limits emissions in each member state and is intended to help states achieve their commitments under the Kyoto Protocol. Many nations with significant greenhouse gas emissions, including the United States, China, and India, had not committed to binding limits on emissions through the Kyoto Protocol or other mechanisms as of the date of this book. However, in December 2007, the Conference of Parties, the supreme body of the Framework Convention, announced the launch of the Bali Action Plan, a comprehensive process that is expected to lead to a decision in 2009 on steps for countries to take on a post-2012 framework.

Instead of adopting limits on emissions, the United States government has addressed climate change with policies that fall into three main categories: (1) programs targeted at enhancing the scientific understanding of climate change, including the Climate Change Science Program (CCSP), directed by the Assistant Secretary of Commerce for Oceans and Atmosphere; (2) programs that support research, development, and deployment of new technologies that could reduce emissions and improve energy efficiency, including the Climate Change Technology Program (CCTP), led by the Department of Energy; and (3) voluntary programs designed to encourage private and public sector entities to curb their greenhouse gas emissions by providing technical assistance, education, and information sharing, including the Environmental Protection Agency's Climate Leaders Program.

The Congress is currently considering various proposals to further address climate change, including actions to mitigate emissions.[6] Mitigation aims to limit the extent of climate change, usually by decreasing greenhouse gas emissions. One possible mitigation policy is an emissions trading program (referred to as a cap-and-trade program). In general, under a cap-and-trade program, such as the European Union's program, the government would limit the overall amount of greenhouse gas emissions from regulated entities. These entities would need to hold allowances for their emissions, and depending on where the regulation was enforced in the economy, each allowance would entitle them to emit 1 ton of carbon dioxide or to have 1 ton of carbon in the fuel they sold. The government could sell the allowances or give them away (or some combination of the two), and establish a market in which the regulated entities could trade the allowances.[7] For example, firms that find ways to reduce their carbon dioxide emissions below their allowed limit could earn revenue by selling their excess allowances to firms that emit more than their limits.[8] In this manner the market would establish a price for a ton of carbon dioxide emissions based on the supply and demand embodied in such trades. Although the program would provide greater certainty that the level of annual emissions would not increase beyond the emissions cap, the cost of the program could vary, depending on factors such as changes in energy prices. Currently, the United States uses a cap-and-trade program to limit pollutants that cause acid rain emitted by electric utilities.

Another possible mitigation policy is a tax on greenhouse gas emissions. A tax would establish a price on emissions by levying a charge on every ton of carbon dioxide emitted, creating an economic incentive for emitters of greenhouse gases to decrease their emissions by, for example, using fossil fuels more efficiently. Unlike a cap-and-trade program, a tax would provide certainty as to the cost of emitting greenhouse gas emissions, but the precise effect of the tax in reducing emissions would depend on the extent to which producers and consumers respond to higher prices.[9]

The Congress is also considering policies that, unlike a cap-and-trade system or a tax, are not based on establishing a market for greenhouse gas emissions. These options include regulatory approaches, such as standards to increase energy efficiency or the use of renewable energy, and nonregulatory approaches, such as investment in research and development of technologies to reduce emissions. The Congress is also considering measures to adapt to climate change, such as developing protective coastal infrastructure to reduce the impact of rising sea levels. The potential benefits and costs associated with a policy, or combination of policies, depend on factors such as their stringency, timing, and effectiveness of stabilizing or reducing greenhouse gas concentrations.

Much of the debate over the direction of U.S. policy to address climate change has centered on the effects that further policy actions could have on economic growth. For decades, economists have sought to inform this debate by analyzing the potential benefits and costs of actions to address climate change. The benefits of such actions could include avoided damages that may result from changing temperatures. For example, many scientists believe slowing the increase in global mean temperatures and the related rise in sea level may limit damage to coastal areas, which are home to the majority of the U.S. population and account for nearly one- third of the gross domestic product.[10] On the other hand, actions to address climate change would impose costs because most emissions stem from the combustion of fossil fuels, which constitute the majority of the nation's energy supply. Thus, actions to mitigate emissions would likely impose higher costs on producers and users of fossil fuels.

To analyze the economic impacts of different policies for addressing climate change, economists have developed sophisticated models that incorporate historical data on the economic effects of changes in energy prices and assumptions about future economic and climatic conditions. These models focus primarily on the benefits and costs of using market- based mechanisms to impose a price on greenhouse gas emissions and generally place a greater emphasis on analyzing the effect on market goods and services, such as fossil fuels, that have readily available prices, than on analyzing the effect on nonmarket goods such as ecological impacts. Key assumptions underlying these models include the degree of international cooperation in mitigating emissions, the rate of technological change, the sensitivity of the climate to changes in emissions, and the degree to which societies adapt to the impacts of climate change.

In using the models to estimate the benefits and costs associated with controlling greenhouse gas emissions, economists have estimated the potential economic effect of establishing a price on greenhouse gas emissions. According to economic theory, the appropriate emissions price should reflect the social costs that result from emissions.[11] For example, in a survey of the economic literature, the IPCC reported that estimates of the damages associated with current greenhouse gas emissions — impacts on public health, ecosystems, and industry

— average about $12 per metric ton of carbon dioxide, with a range from $3 to $95 per ton (2005 dollars).[12] The wide range of estimates primarily reflects differences in the models used and key assumptions.

In this context, you asked us to elicit the opinions of experts in the field of climate change economics on (1) actions the Congress might consider to address climate change and what is known about the potential benefits, costs, and related uncertainties of these actions and (2) the key strengths and limitations of policies or actions to address climate change. To respond to these objectives, we collaborated with the National Academy of Sciences (NAS) to identify and recruit experts with experience analyzing the economic effects of climate change policies. NAS recruited 25 experts affiliated with U.S.-based institutions who have conducted research on the benefits, costs, or uncertainties associated with actions to address climate change, and with in-depth experience in assessing the economic impacts and trade-offs of climate change policies. The experts who served on the panel represent the range of existing research on the economics of climate change, with expertise in areas such as environmental, natural resource, and agricultural economics, and some have served as advisors to the United States government, including as members of the Council of Economic Advisors under the current and former administrations.

To address the first and second objectives, we (1) reviewed relevant climate change academic literature and documents developed by federal agencies and (2) met with agency officials from the Environmental Protection Agency (EPA); the Department of Energy (DOE), including the Energy Information Administration (EIA); the Department of Commerce, including the National Oceanic and Atmospheric Administration (NOAA); the United States Department of Agriculture (USDA); and the National Aeronautics and Space Administration (NASA). To structure our questions and gather opinions from the panelists on our objectives, we used a modified Delphi method, an iterative and controlled feedback approach. With two Web-based questionnaires, we first gathered opinions from the panel on the key topics, and used their responses to develop the second round of questions. In the second round, panelists reacted to the issues and topics discussed in the first round, answering primarily closed-ended questions. We used this approach to eliminate the potential bias associated with live group discussions, and to incorporate more panelists than a live panel would allow. Including more panelists also enabled us to obtain the broadest possible range of opinion. Of the 25 panelists NAS recruited to participate, 21 agreed and were sent the first questionnaire. Nineteen responded to the first questionnaire, and 18 responded to the second. After the responses from the second round were compiled, the panelists were given 2 weeks to comment on a summary of the results. In addition, we followed up with several panelists to verify their responses and elaborate on certain topics. The information presented in this book is primarily from the second questionnaire and represents the views of the 18 experts who participated in both rounds and not GAO's (See app. I for a more detailed description of our scope and methodology). We conducted our work from September 2006 to May 2008.

Chapter 1

RESULTS IN BRIEF[*]

All of the economists on the panel agreed that the Congress should consider establishing a price on greenhouse gas emissions using a market-based mechanism but expressed differing views on the type of mechanism and its stringency. In addition, a majority of panelists recommended implementing a portfolio of actions, including at least one complementary policy action in areas such as research and development, adaptation, or international negotiations and assistance. In terms of the mechanism for establishing a price, 8 panelists preferred a cap-and-trade system with the government having the ability to use a cost control mechanism (called a safety valve) if the price of permits exceeds certain levels, 7 preferred a tax on emissions, and 3 preferred a cap-and-trade system without the safety valve. Despite key uncertainties associated with estimating potential costs and benefits, 14 of the 18 panelists said they were at least moderately certain that the benefits of their preferred portfolio of actions would outweigh the costs, and 4 did not respond to questions on this topic. With respect to the stringency of their recommended market-based mechanism, 7 panelists said the price per ton of emissions should range from less than $1 to $10, 7 said from $11 to $20, and 3 said it should exceed $20 (2007 dollars).[13] In addition, the majority of panelists said that the price on emissions should gradually increase over time. Further, all of the panelists said that the price should be implemented by 2015, and that it should apply to all sectors of the economy.

Panelists identified general categories of potential benefits and costs and provided some quantitative estimates or cited estimates from related studies. The

---

[*] Excerpted from GAO Report GAO-08-605, dated May 2008.

panelists rated avoided climate change damages as the most important type of benefits; these damages may include flooding from rising sea levels and extreme weather events. Some panelists also provided cost estimates for their preferred actions and cited key studies underlying those estimates. Overall, the panelists said that estimates of costs are more useful for informing congressional decision makers than estimates of benefits, but all of the panelists that responded to the applicable questions on this topic said that the estimates of costs and benefits from integrated assessment models were at least somewhat useful. The panelists identified key uncertainties that affect these estimates, including uncertainty about the extent to which rising temperatures could lead to abrupt changes in the climate system, the science of climate change, and the potential economic effects of actions to address climate change. Despite these uncertainties, 16 of the 18 panelists agreed that the United States should limit emissions as soon as possible, regardless of the efforts of other nations to adopt similar policies. At the same time, the majority of the panelists said that it was at least somewhat important to participate in international negotiations, either to facilitate climate agreements or to enhance the credibility or influence of the United States.

Panelists identified key strengths and weaknesses of alternative policy approaches that should be of assistance to the Congress in weighing the potential benefits and costs of different policies for addressing climate change. Notably, the experts discussed the greater certainty of attaining emissions targets under a cap-and-trade system versus the greater efficiency of a tax in achieving emissions reductions at a lower cost. On average, they rated cost-effectiveness as the most important criterion for evaluating various policy options, and used it and other important criteria to compare the strengths and limitations of different actions to address climate change. Some panelists said that a cap-and-trade program would be more effective in achieving a desired level of greenhouse gas emissions because, unlike a tax, it would provide certainty that emissions would not exceed a certain level. However, some of the panelists also said that taxes were more economically efficient than a cap-and-trade program because the price of emissions would be certain and would not be susceptible to market fluctuations that could lead to increased costs. In addition, some panelists felt a cap could be more administratively burdensome than a tax.

Nonetheless, some of the panelists that preferred a tax said that a cap-and-trade program, especially if it included cost-minimizing components, would be an acceptable second option to address climate change. Eight panelists preferred a hybrid approach where the government would create a cap-and-trade system with the option of selling additional permits if the market price of permits exceeded a certain level, thereby minimizing risks of adverse economic consequences.

However, the emissions reductions achieved by a hybrid program would be less certain than a standard capand-trade program if the price control mechanisms came into effect. As for distributing emissions permits under a cap-and-trade or hybrid program, the majority of panelists favored at least partial auctioning of permits rather than free distribution. They noted that the government could redistribute revenue from permit auctions to offset adverse effects on consumers or particular sectors of the economy. The panelists also discussed the strengths and limitations of other policy options, including research and development of technologies, adaptation to the anticipated impacts of climate change, revising energy efficiency standards, and reforming subsidies for fossil fuel production and other industries.

Chapter 2

BACKGROUND

Greenhouse gases — including carbon dioxide, methane, nitrous oxide, and other substances — trap a portion of the sun's heat in the atmosphere and prevent the heat from returning to space. The insulating effect, known as the greenhouse effect, moderates atmospheric temperatures, keeping the earth warm enough to support life. According to the Intergovernmental Panel on Climate Change (IPCC), global atmospheric concentrations of these greenhouse gases have increased markedly as a result of human activities over the past 200 years, contributing to a warming of the earth's climate.[14]

The IPCC generally attributes increases in average global air and ocean temperatures, widespread melting of snow and ice, and rising mean global sea levels to a warming of the earth's climate system. Furthermore, according to the IPCC, the oceans have absorbed more than 80 percent of the heat added to the earth's climate system, causing seawater to expand, thereby contributing to sea level rise. Scientists have also reported that mountain glaciers and snow cover have declined, on average, in both hemispheres, and that widespread decreases in the sizes of glaciers and polar ice caps, combined with losses in the ice sheets of Greenland and Antarctica very likely contributed to a sea level rise of 0.17 meters during the 20th century.

The effect of increases in atmospheric concentrations of greenhouse gases and temperature on ecosystems and economic growth is expected to vary across regions, countries, and economic sectors (see Table 1).

Table 1. Potential Impacts of Climate Change by Sector

Sector	Major projected impacts
Agriculture, forestry, and ecosystems	– Increased yields in colder environments
	– Decreased yields in warmer environments
	– Increased insect outbreaks
	– Increased danger of wildfires
	– Damage to crops
	– Waterlogging of soils
	– Land degradation
	– Increased livestock deaths
	– Uprooting of trees
	– Damage to coral reefs
	– Salinization of irrigation water, estuaries, and freshwater systems
Water resources	– Effects on some water resources, such as increased salinization of groundwater and decreased availability of freshwater for humans and ecosystems
	– Increased water demand
	– Water quality problems
	– Adverse effects on quality of surface and groundwater
	– More widespread water scarcity
	– Power outages causing disruption of public water supply
	– Decreased freshwater availability due to saltwater intrusion
Human health	– Reduced human mortality from decreased cold exposure
	– Increased risk of heat-related mortality
	– Increased risk of deaths, injuries, and infectious respiratory and skin diseases
	– Increased risk of food and water shortage
	– Increased risk of malnutrition
	– Increased risk of water- and food-borne diseases

Industry, settlement, and society	– Increased risk of deaths and injuries by drowning and floods
	– Reduced energy demand for heating
	– Increased energy demand for cooling
	– Declining air quality in cities
	– Reduced disruption to transport due to snow, ice
	– Disruption of settlements, commerce, transport, and societies due to flooding
	– Pressures on urban and rural infrastructures
	– Water shortages for settlements, industry, and societies
	– Reduced hydropower generation potential
	– Potential for population migration
	– Disruption by flood and high winds
	– Withdrawal of risk coverage in vulnerable areas by private insurers
	– Costs of coastal protection versus costs of land use relocation
	– Potential for movement of populations and infrastructure

Source: IPCC, Working Group III, AR4, Summary for Policymakers.

For example, small island nations are particularly at risk because of their vulnerability to sea level rise, poor coastal infrastructure, and economies that rely heavily on coastal fishing and tourism. Alternatively, while certain areas of the United States may be adversely affected by rising sea levels, its diverse economy, significant resources, and established infrastructure may help moderate the negative effects associated with climate change. Figure 1 shows a selection of projected impacts from climate change on different regions, assuming that greenhouse gas emissions and concentrations continue to increase at current rates.

According to the IPCC, in 2004, developed countries, including the United States, constituted 20 percent of global population, but were responsible for nearly half of global greenhouse gas emissions.[15] However, in the absence of mitigation policies, the IPCC projects that between 2000 and 2030, two-thirds to three-quarters of the projected increase in global carbon dioxide emissions will occur in developing countries.[16]

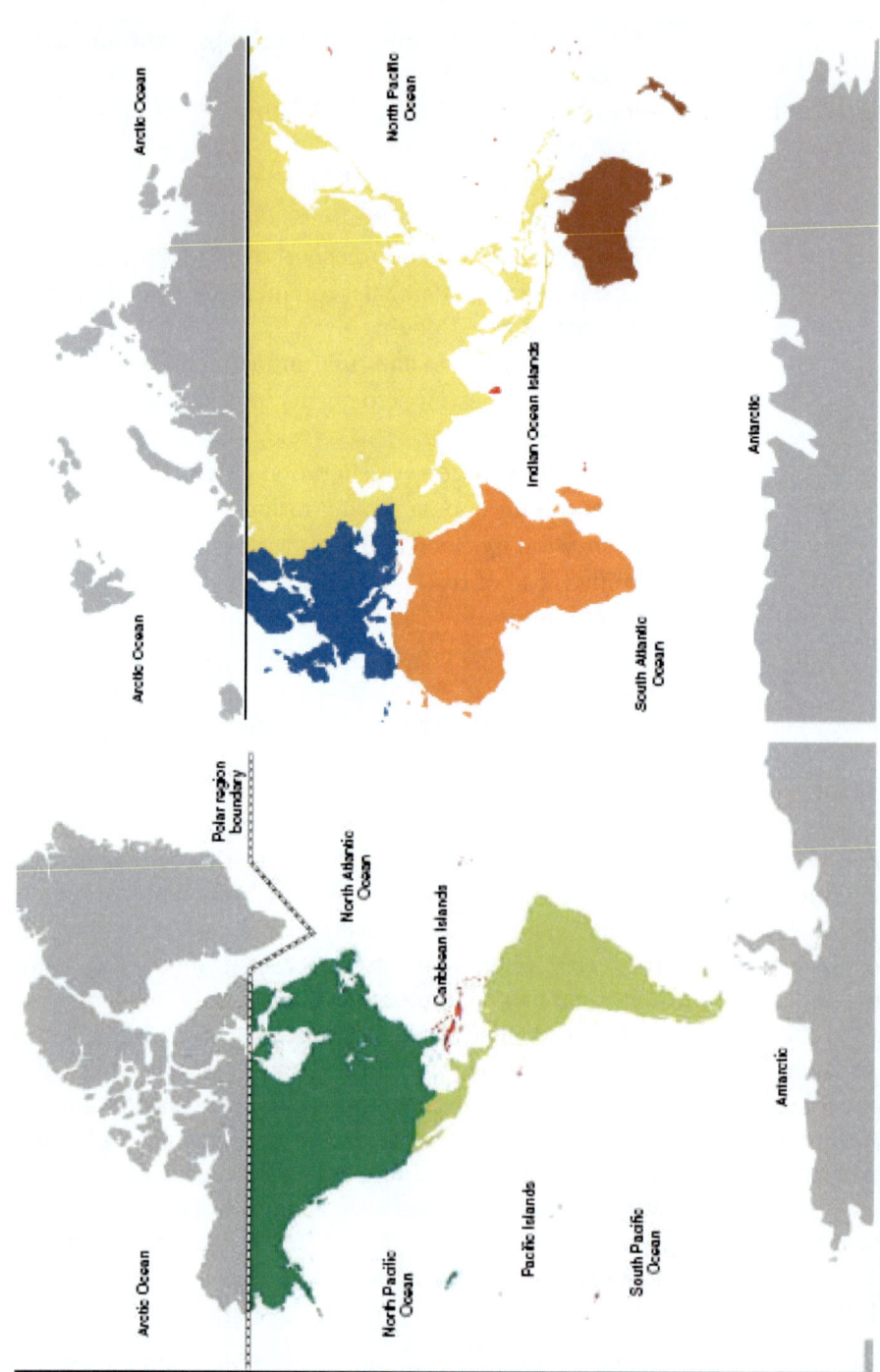

Source: GAO and Map Resources based on IPCC analysis.

North America:
- Decreased snowpack, more winter flooding, and reduced summer flows, exacerbating competition for already limited water resources in western mountains
- Extended periods of high fire risk and large increases in area burned in forest areas
- Increased aggregate yields of rain-fed agriculture, but with variability among regions
- Increased heatwaves with greater intensity and frequency in cities that currently experience heatwaves
- Increased vulnerability to climate variability and future climate change due to population growth and the rising value of infrastructure in coastal areas, with current adaption uneven and readiness for increased exposure low

Latin America:
- Gradual replacement of tropical forest by savanna in eastern Amazonia (by midcentury)
- Significant biodiversity loss in many areas of tropical Latin America
- Decreased agricultural productivity, with adverse consequences for food security in drier areas
- Increased flooding in low-lying areas and increasing sea surface temperatures, adversely affecting coral reefs and causing shifts in the location of southeastern Pacific fish stocks
- Decreased water availability for human consumption, agriculture, and energy generation
- Effectiveness of adaptation efforts outweighed by the lack of capacity building and appropriate political, institutional, and technological frameworks

Polar regions:
- Reduced thickness and extent of glaciers and ice sheets and changes in natural ecosystems
- Detrimental impacts on infrastructure and traditional indigenous ways of life in the Arctic
- Beneficial impacts including reduced heating costs and more navigable northern sea routes
- Increased vulnerability for specific ecosystems and habitats in both polar regions as climate barriers to species invasions are lowered

Small islands:
- Adverse impacts on local resources, such as fisheries, reducing the value of these destinations for tourism
- Increased flooding, storm surge, erosion, and other coastal hazards, threatening vital infrastructure, settlements, and facilities of island communities
- Substantially decreased freshwater availability
- Increased invasion by non-native species, particularly on

Africa:
- Extensive increases in water scarcity (by 2020)

Figure 1. (Continued)

- Severely compromised agricultural production (by 2020)
- Negative effects on local food supplies
- Projected sea-level rise will affect heavily populated low-lying coastal areas (by 2100)
- High vulnerability to impacts due to low adaptive capability

Asia:
- Increased flooding, avalanches, and water scarcity in the Himalayas
- Increased freshwater scarcity in Central, South, East, and Southeast Asia (by 2050s)
- Increased risk of flooding in heavily populated megadeltas in South, East, and Southeast Asia
- Increased crop yields in East and Southeast Asia will be offset by decreased crop yield in Central and South Asia; coupled with rapid urbanization and population growth, the risk of hunger is expected to remain very high in several developing countries
- Increased health risks associated with flooding, droughts, and increasing coastal water temperatures

Australia and New Zealand:
  - Increased water scarcity in southern and eastern Australia and New Zealand (by 2030)
- Significant biodiversity loss in ecologically rich sites such as the Great Barrier Reef (by 2020)
- Increased risk to coastal development and population growth due to sea-level rise and increases in the severity and frequency of storms and coastal flooding (by 2050)
- Decreased agricultural and forestry productivity for much of southern and eastern Australia and parts of eastern New Zealand
- Longer growing seasons in western and southern New Zealand

Europe:
- Increased risk of inland flooding, and coastal flooding and increased coastal erosion
- Glacier retreat, reduced snow cover and winter tourism, and extensive species loss in mountainous areas (by 2080)
- Higher temperatures, increased drought, reduced water availability and crop productivity, increased health risks, and increased frequency of wild fires in Southern Europe
- Decreased summer precipitation, reduced water availability, increased health risks, reduced forest production, and increased frequency of peatland fires in Central and Eastern Europe
- Decreased demand for heating, increased crop yields, and forest growth in Northern Europe outweighed by increased frequency of winter floods and endangered ecosystems

Figure 1. Figure 1: Potential Impacts of Climate Change by Geographic Region.

MITIGATING GREENHOUSE GAS EMISSIONS

Many developed countries have begun to mitigate or reduce greenhouse gas emissions by adopting policies such as carbon taxes, cap-and-trade programs, energy efficiency standards, financial incentives (e.g., subsidies or tax credits), voluntary agreements, education campaigns, and research, development, and deployment of advanced technologies.[17] For example, the European Union Emission Trading Scheme (EU ETS) is a cap-andtrade system in which energy-intensive industries in the European Union buy or sell emission allowances to help meet member states' commitments under the Kyoto Protocol. The EU ETS covers over 11,000 energy- intensive installations, such as oil refineries and steel plants, in 25 member countries and covering nearly half of Europe's carbon dioxide emissions. Governments may also use a portfolio of policies; for example, a cap-andtrade system may be pursued in combination with energy efficiency standards and financial incentives for certain sectors. Table 2 shows selected policies and instruments that have been shown to be environmentally effective at a national level.

In addition, carbon capture and storage can supplement other climate change mitigation policies. Carbon capture and storage involves separating and storing carbon dioxide from an industrial or energy-related source, thereby preventing carbon dioxide emissions into the atmosphere. Carbon capture and storage is most commonly used to enhance oil and gas recovery in depleted fields. Efforts to capture carbon dioxide from power generation or industrial processes are currently the main focus of research and development of the technology. This process has the potential to reduce emissions, but its widespread use may be limited by several barriers, including technological feasibility, costs, regulatory issues, and environmental concerns.

Another option is to allow for carbon offsets, which is a way for consumers and producers to compensate for greenhouse gas emissions occurring in one location by reducing or avoiding emissions somewhere else. For example, a manufacturing facility in the United States could compensate for its emissions by purchasing carbon offsets from a tree- planting project in South America. Carbon offsets are traded in compliance markets, to satisfy requirements to limit emissions, and in voluntary markets, where emissions reductions are not required but may serve other purposes. For example, carbon offsets serve as a mechanism for complying with the emissions reduction requirements of the EU ETS. Under this scheme, certain regulated entities may choose to comply with emissions limits by purchasing offsets rather than by reducing their own emissions.

Table 2. Selected Policies, Measures, and Instruments Currently Used by Various Nations to Address Climate Change

Sectr	Plicies, measures, and instruments
Energy Supply	− Reduction of fossil fuel subsidies − Taxes or carbon charges on fossil fuels − Incentives for renewable energy − Public research, develpment, and deplyment of low-emission technologies
Transprtatin	− Mandatory fuel ecnomoy standards, biofuel blending, and Co_2 standards for road transport − Taxes on vehicle purchase, registration, and use, and on motor fuels, road and parking pricing − Land use regulations and infrastructure planning − Investment in public transport facilities and nonomotorized forms of transport − Public research, develpment, and deplyment investment in lw-emissin technlgies
Buildings	− Appliance standards and labeling − Building codes and certification − Demand-side management prgrams to incentivize customers to purchase energy-efficient prducts − Public sector leadership programs, including procurement requirements for gvernments − Incentives for energy service companies − Public research, development, and deployment investment in low-emission technologies
Industry	− Performance standards − Subsidies, tax credits − Tradable permits − Voluntary agreements − Public research, development, and deployment investment in low-emission technologies
Agriculture	− Financial incentives and regulations for improved land management, maintaining soil carbon content, efficient use of fertilizers and irrigation − Public research, development, and deployment investment in low-emission technologies
Frestry	− Financial incentives (national and international) t increase forest area, to reduce deforestation, and to maintain and manage forests − Land use regulatoin and enforcement

	– Public research, development, and deployment investment in low-emission technlgies
Waste management	– Financial incentives for improved waste and wastewater management
	– Renewable energy incentives or obligations
	– Waste management regulations
	– Public research, development, and deployment investment in low emission technologies

Source: IPCC, Working Group III, AR4, Summary for Policymakers.

ADAPTATION POLICIES TO REDUCE VULNERABILITY TO CLIMATE CHANGE

In addition to mitigating greenhouse gas emissions, policies to adapt to climate change could help reduce the vulnerability of countries and regions to potentially adverse impacts. For example, raising river or coastal dikes could protect coastal communities and resources from sea level rise. The vulnerability of a country or region depends on both on the susceptibility of a political, economic, or natural system to the adverse effects of climate change and the capacity of a society to adjust to the expected change. For example, less developed economies may face difficulty adapting to climate change because of poor infrastructure, poverty, and resource constraints. Adaptation may be viewed as a risk-management strategy for protecting vulnerable countries, sectors, and communities that might be affected by changes in the climate and related impacts.

In December 2007, members of the Conference of the Parties to the Framework Convention agreed to launch a comprehensive process that will lead to the adoption of a decision in 2009 on next steps for countries to take on climate change. This process, called the Bali Action Plan, includes provisions that require signatories, including the United States, to undertake efforts to enhance international cooperation on adaptation, such as vulnerability assessments; capacity building; risk management; and reduction strategies; and the integration of adaptation into planning decisions.[18]

Estimating the Potential Benefits and Costs of Actions to Address Climate Change

Economists and other researchers have developed integrated assessment models that utilize economic and climate and other environmental data to assess the economic consequences associated with different policies for addressing climate change. These models vary in structure and scope, but generally include historical data on the U.S. and international economies, emissions and atmospheric concentrations of greenhouse gases, and global temperature. Typically, economists use the models to estimate the economic consequences of policy actions, by comparing the present values of the economic costs of an action such as tax or cap and trade and the future benefits it would be expected to generate, relative to a businessas-usual emissions projection (for example, no significant reductions in greenhouse gas emissions). In judging whether a policy action would be preferable, economists evaluate policies using criteria such as (1) economic efficiency, where the action maximizes potential net benefits (total benefits minus total costs), compared to business as usual and (2) cost-effectiveness, where the action achieves the chosen policy objective, such as an emissions reduction target, at least cost.

In general, the economic costs that society would incur from taking action to address climate change will begin to occur immediately, while the economic and environmental benefits will mainly occur decades in the future as atmospheric concentrations of greenhouse gases and global temperatures stabilize. The economic costs of taking action represent the value of the goods and services that society would forgo to allocate resources to the emissions control policy, including compliance expenditures, administration and enforcement costs, and other costs that the action might impose on the economy (for example, as a result of higher prices). For example, because the energy sector and energy- intensive industries generate substantial emissions of greenhouse gases, a significant component of the potential cost associated with actions to address climate change relates to the impacts of changes in energy prices. The economic costs of reducing emissions to stabilize atmospheric greenhouse gas concentrations will increase as the stringency of emissions reduction goals increase, and correspondingly, as the stabilization goal decreases.

The potential economic benefits of policies to address climate change generally consist of the effects of stabilizing or reducing the atmospheric concentration of greenhouse gases and global temperature on human welfare. Typically, benefits are measured in terms of the damages that would be averted if an action were taken to address climate change. For example, under a business-as-usual scenario, researchers project that further increases in atmospheric concentrations of greenhouse gases and global temperature could reduce agricultural productivity in certain parts of the world, increase the incidence of diseases in certain climates, and reduce the environmental goods and services provided by some ecosystems. In addition, researchers have estimated that limiting greenhouse gas emissions and slowing the increase in atmospheric concentrations and global temperature would avoid some damages.

Estimating the potential benefits associated with actions to address climate change, however, can pose challenges, partly because of uncertainty about the magnitude of climate impacts and the resulting effect on human welfare. In particular, scientists face challenges estimating the effects of climate change at the regional and local level. In addition, rising global temperature could involve unexpectedly abrupt changes in the climate, which would be more costly than if changes are more moderate. According to NAS, global warming and other human alterations of the earth's climate system may increase the possibility of large and abrupt regional or global climatic events. However, NAS concluded that because the abrupt climate changes of the past are not yet fully explained, future abrupt changes cannot be predicted with any confidence and climate surprises are to be expected.[19] In addition, partly because many of the environmental goods and services expected to be affected by climate change are generally not bought and sold in markets, it is difficult to develop reliable estimates of the value associated with the damages expected from climate change.[20]

Furthermore, estimating the economic consequences of various actions to address climate change involves consideration of the fact that the benefits and costs will occur in different time periods and should be expressed in comparable present value terms. For example, because greenhouse gases accumulate in the atmosphere over very long periods of time, the effect of current actions to stabilize concentrations and global temperature might take decades to manifest themselves. As a result, the future benefits and costs of actions to address climate change are typically discounted (by using a discount or interest rate) to estimate their present value. In general, discounting is used to reflect the extent to which individuals trade off current for future consumption. Nonetheless, because discounting generally attaches a lower weight to future impacts compared to near-term impacts, the choice of the discount rate is an important factor in assessing the economic consequences of actions to address climate change.[21]

Moreover, although the cost of any U.S. action to address climate change will be primarily borne by U.S. producers and consumers, the economic benefits associated with the action would also accrue to other countries that face greater vulnerability to climate change. According to the IPCC, the global distribution of climate change impacts varies throughout the world and nonclimate stresses such as poverty and food insecurity can increase a country's vulnerability to climate change by reducing its resilience and capacity to adapt. For example, the IPCC estimated that Africa could face increased risk of water scarcity and reduced food security.[22] However, because African nations account for a negligible share of total greenhouse gas emissions, it is expected that they will bear a much smaller portion of the overall costs or responsibility for greenhouse gas mitigation.

CLIMATE CHANGE POLICIES CURRENTLY UNDER CONSIDERATION BY THE CONGRESS

The eight climate change mitigation bills currently under consideration by the Congress provide an overview of the potential for simultaneously pursuing a portfolio of actions. All eight bills include provisions for a capand-trade system in combination with initiatives to promote the development and adoption of low-carbon technologies.[23] Also, three bills would require that a specific amount of electricity be generated by renewable energy — generally called a Renewable Portfolio Standard — including wind and solar energy, and energy efficiency performance standards. Five bills would apply the emissions caps to specific sectors of the economy, such as electricity, transportation, and industry, while caps under the other three would not be limited to specific economic sectors.

Chapter 3

DESPITE SOME UNCERTAINTY REGARDING THE POTENTIAL ECONOMIC IMPACT, ALL OF THE PANELISTS SUPPORTED ESTABLISHING A PRICE ON GREENHOUSE GAS EMISSIONS

All of the panelists agreed that the Congress should consider establishing a price on greenhouse gas emissions using a market-based mechanism, but they expressed differing views on the type of mechanism and its stringency. In addition, 14 of the 18 panelists were at least moderately certain that the benefits of their suggested portfolio of actions would outweigh the costs.[24] Most of the panelists identified either a tax on emissions or a cap-and-trade program with a safety valve as the preferred mechanism to establish a price on emissions, and the majority believed a portfolio of additional actions to address climate change could complement the market-based mechanism. Some panelists also identified general categories of benefits and costs associated with their recommended actions, and rated the usefulness of benefit and cost estimates derived from integrated assessment models. Overall the panel rated estimates of costs as more useful than estimates of benefits for informing congressional decision making, with some panelists citing uncertainties associated with the future impacts of climate change as a limitation to estimating benefits. Finally, while some panelists said that the United States should proceed cautiously if it acts unilaterally, the majority of panelists agreed that it should establish a price on greenhouse gas emissions as soon as possible regardless of the extent to which other countries adopt similar policies.

ALL OF THE PANELISTS AGREED THAT THE CONGRESS SHOULD CONSIDER ESTABLISHING A PRICE ON GREENHOUSE GAS EMISSIONS, AND THE MAJORITY RECOMMENDED COMPLEMENTARY POLICIES

All of the panelists agreed that the Congress should consider a market-based mechanism to establish a price on greenhouse gas emissions and supported implementation of the policy by 2015. Opinions varied on whether the Congress should implement a cap-and-trade system or a tax to control greenhouse gas emissions, with eight panelists preferring a capand-trade program with a safety valve (sometimes referred to as a hybrid system), seven preferring a tax, and three preferring a cap-and-trade program. All of the panelists agreed that the policy should target all sectors of the economy, and the majority believed that it should include all greenhouse gases. For example, one panelist stated that by establishing a price on emissions from all sources in the United States with no exceptions, the policy would equilibrate the marginal cost of reducing emissions across all sources, making it economically efficient.

The panelists varied in their views on the stringency of the market-based regulatory mechanism that they supported to place a price on greenhouse gas emissions.[25] For example, in proposing an initial price on emissions, seven panelists said it should range from less than $1 to $10, six said from $11 to $20, and four said it should be greater than $20 (2007 dollars per metric ton of carbon dioxide equivalent).[26] In addition, while most panelists said the price should increase over time, they varied in their views on the preferred rate of increase. For example, some panelists provided estimates ranging from 2 percent to 5 percent per year adjusted for inflation, while another panelist said more generally that it should be reevaluated periodically (for example, every 5 years) and rise as marginal damages of climate change rise. Some panelists noted the importance of a long-term commitment to establish a price on emissions and the flexibility to adjust the price and rate of increase as new information becomes available. For example, one panelist stated that certainty in setting emissions reductions goals was necessary for firms that would have to make substantial investments in new emissions reduction technologies.

Another panelist added that while the program should run into the distant future, it should include an explicit feature to revise taxes or emissions targets based on near-term mitigation experience, new research on climate change science and impacts, and actions by other countries.

Under a hybrid approach, the government would establish a cap-and-trade system with a safety valve, a mechanism under which the government would sell additional permits if the market price exceeded a predetermined level, which would represent the maximum permit price and thus an upper limit on control costs. A safety valve price would be established, which would represent the maximum permit price and thus an upper limit on control costs. Panelists preferring a hybrid policy varied in their recommendations for an initial safety valve price, with five preferring a safety valve price identical or close to their targeted initial market price for greenhouse gas emissions and two recommending a price moderately or well above their recommended market price (see app. II for additional information on this topic). For example, a panelist recommending a targeted initial market price of $0.55 per metric ton of emissions (in carbon dioxide equivalent) recommended an identical safety valve price, adding that the price should not be much higher than the targeted marginal cost of reducing emissions. Another panelist recommending an initial market price between $11 and $20 said that the safety valve should depend on the stringency of the policy. When providing an example, the panelist added that if the initial market price per metric ton of emissions was $20, then the safety valve should be set three to four times above the initial price of emissions, or between $60 and $80 (See app. II for additional information on this topic).

In addition, seven of the eight panelists that prefer a hybrid policy believe that the price of the safety valve should increase over time, and the majority of panelists said that the safety valve should be reevaluated periodically based on new information. For example, one panelist said that the safety valve price should increase by 2.5 percent plus inflation annually, and that the government should reevaluate the stringency of the policy every 5 to 10 years based on new information. Another panelist said that the safety valve provision could be abandoned altogether after a periodic review of the adequacy of the policy and its costs.

Overall, eight of the panelists said that a market-based mechanism should be imposed "upstream," where fossil fuels first enter the economy, four preferred a "downstream" mechanism that regulated direct and indirect emitters, and five preferred a mechanism with both upstream and downstream components. An upstream system would require fossil fuel producers, such as extractors and processors, and importers, to pay a tax or to hold permits based on the carbon content of the fuels. Alternatively, a downstream system would regulate sources such as electric utilities that combust fossil fuels and emit greenhouse gases. While environmental effectiveness would likely be the same under either approach, one panelist said that an upstream policy would have lower administrative costs when compared to a downstream system partly because it involves a much more manageable set of firms. Of the panelists who said they preferred a combination of upstream and downstream provisions, three preferred a policy that had downstream provisions for carbon capture and storage from utilities or other processes that remove carbon from the atmosphere. See appendix II for additional information on experts' preferred design for a market-based policy.

In the second round questionnaire, we asked panelists to rate the importance of complementary actions to address climate change that had been recommended in the first round. We also asked panelists to identify their recommended portfolio of actions for the Congress to consider. When rating the importance of recommended actions, the panelists gave the highest average ratings to funding of research, development, and deployment of zero-carbon and low-carbon technologies, and participation in international negotiations (see fig. 2 and app. III for more detail). For example, recognizing that the private sector may not invest in some technologies, one panelist recommended public investment to accelerate development of carbon capture and storage from electricity generation. This panelist also advocated expanded research into climate change adaptation in the southern United States and in developing countries for agriculture, forestry, and fisheries, emphasizing the importance of preparing for inevitable climate impacts. Another panelist recommended that the United States engage in international negotiations to make similar emissions reductions commitments, and added that the United States should consider helping other countries meet their targets as a part of international aid.

When recommending a portfolio of policy options for the Congress to consider, 14 of the 18 panelists identified actions to address climate change in addition to placing a price on greenhouse gas emissions (see app II, table 4, for more detail). Of the 14 holding this view, 10 said research and development in low- or zero-carbon technologies or research in the basic science of climate change should be part of the portfolio, 7 said international negotiations or assistance to developing countries should be included, and 6 said adaptation should factor into a portfolio of actions (see app. II, table 4, for more detail). When asked how certain they were that their recommended portfolio of actions to address climate change were economically justified, 14 of the 18 panelists were at least moderately certain that the benefits of their suggested actions would outweigh the costs. The remaining 4 panelists did not know or did not provide a response to this question.

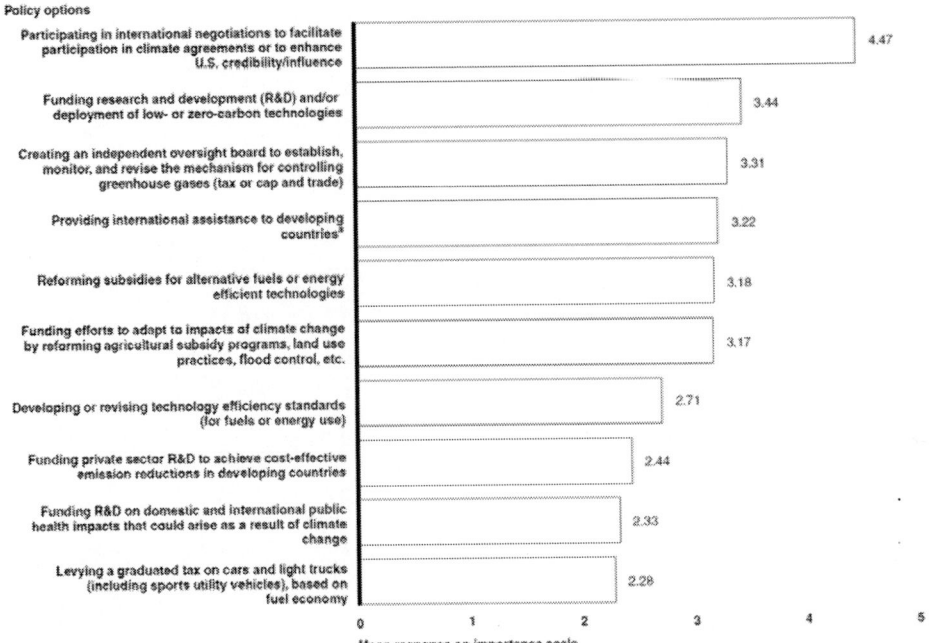

1 Not at all important
2 Somewhat important
3 Moderately important
4 Quite important
5 Extremely important.

Note: Total responses for each question may range from 15 to 18. See appendix III for additional details.

[a] Providing international assistance would include compensation for climate change impacts or assistance deploying low-carbon technologies.

Figure 2. Mean Panelist Ratings of the Importance of Additional Policy Options to Address Climate Change.

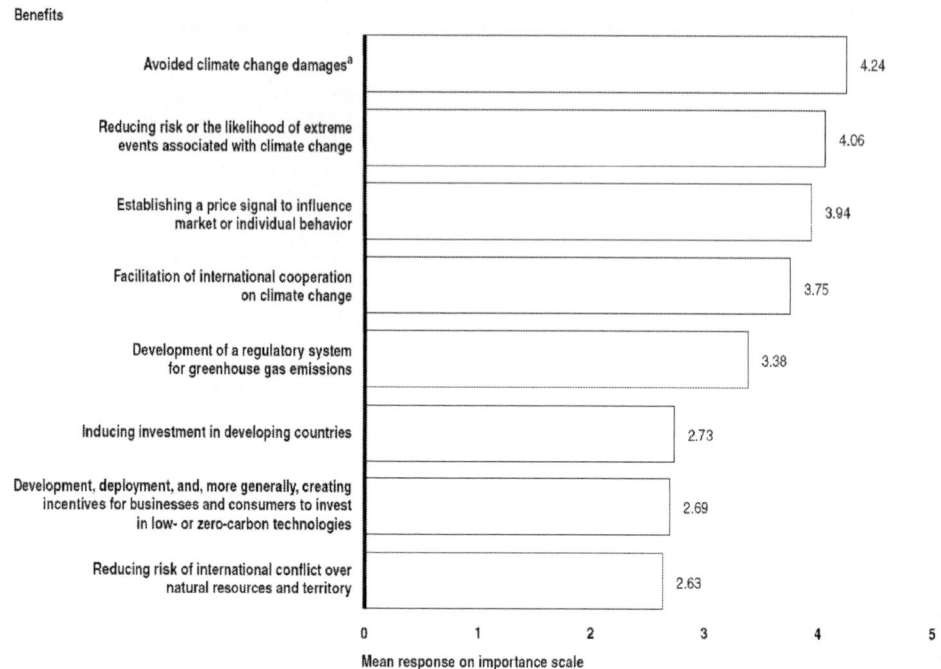

1 Not at all important
2 Somewhat important
3 Moderately important
4 Quite important
5 Extremely important

Note: Total responses for each question may range from 15 to 18. See appendix III for additional details.

[a]Avoided climate change damages includes flooding, extreme weather events, crop damage, impacts on sensitive ecosystems, public health, and species loss.

Figure 3. Mean Panelist Ratings of the Importance of General Categories of Potential Benefits as a Rationale for Addressing Climate Change.

PANELISTS DESCRIBED THE POTENTIAL BENEFITS AND COSTS OF ACTIONS TO ADDRESS CLIMATE CHANGE

In the first questionnaire, we asked panelists to identify potential categories of costs and benefits associated with actions to address climate change. In the second questionnaire, we asked them to rate the importance of categories of benefits as a rationale for addressing climate change.[27] On average, panelists rated avoiding damages such as those from flooding, impacts on sensitive ecosystems, public health, and species loss as the most important category of potential benefits (see fig. 3 and app. III for more detail). In addition, panelists rated reducing risk of extreme or irreversible climate events as the second most important category of potential benefits. For example, one panelist discussed the benefits of reducing the probability of abrupt or catastrophic climate events such as dramatic sea level rise by stabilizing atmospheric concentrations of greenhouse gases. Another noted the importance of avoiding damages by reducing the risks of vulnerability to water scarcity, hunger, or the frequency of storm events.

The panelists also rated establishing a price signal to influence market or individual behavior and facilitating international cooperation on climate change as an important category of benefits that could serve as a rationale for actions to address climate change. For example, one panelist stated that the United States needs an unavoidable price signal to fully harness the innovativeness of the U.S. industrial and scientific communities and to provide incentives to reduce emissions. Another panelist stated that a modest near-term market-based policy could provide an opportunity to learn about the effectiveness of an emissions tax or cap-and-trade program, and stimulate research to inform the stringency of future policies.

In the first questionnaire, some panelists presented estimates of the costs that would be associated with their proposed actions to address climate change. Panelists' estimates of the impacts on social welfare, including the effects on economic growth, varied depending on the type and stringency of the policy recommended. Some panelists cited their own research and other academic studies, as well as several assessment reports by domestic and international governmental entities, as credible estimates of the economic costs associated with their proposed actions, but some noted that policy choice and stringency can have a large impact on the cost estimates. For example, one panelist cited a recent modeling effort by the U.S. Climate Change Science Program (CCSP) that projected that by 2060, gross world product (GWP) would decline by between 0 and 6.7 percent per year under emissions reduction scenarios that achieved different levels of atmospheric stabilization of greenhouse gases.[28] However, another panelist stated that welfare cost estimates from the study could be reduced by as much as 50 percent by incorporating certain policy features such as carbon offsets into the study's policy scenario. Further, another panelist added that it is difficult to provide a comprehensive list of potential costs given the breadth of possible policy scenarios.

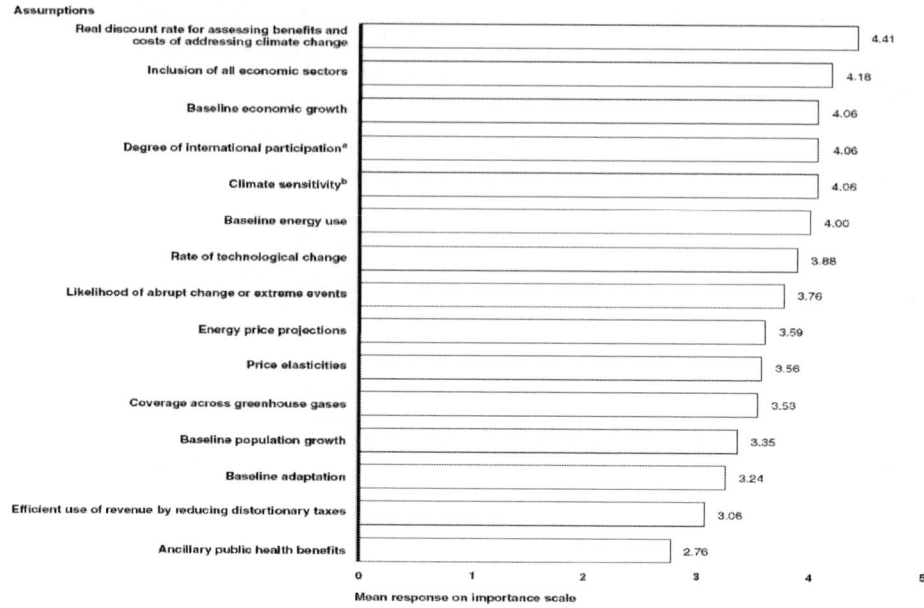

1 Not at all important
2 Somewhat important
3 Moderately important
4 Quite important
5 Extremely important

[a] Degree of international participation means how quickly and to what extent high-emitting nations implement an emissions reduction policy.

[b] Climate sensitivity is the change in global mean temperature that results when the climate system attains a new equilibrium as the result of a doubling in the atmospheric carbon dioxide concentration.

Source: GAO analysis.

Note: Total responses for each question may range from 15 to 18. See appendix III for additional details.

Figure 4. Mean Panelist Ratings of the Importance of Types of Assumptions in Integrated Assessment Models.

Some panelists also commented on the effect of their suggested actions on energy prices. For example, one panelist said that a mandatory emissions reduction program that establishes a price around $20 per ton of carbon dioxide (2005 dollars) would increase gasoline prices by approximately 20 cents per gallon and residential electricity prices by approximately 1 cent per kilowatt hour above business as usual estimates. Another expert estimated that a $20 price per ton of carbon dioxide equivalent would result in a 10 percent increase in consumer energy prices. Both experts cited a recent study by the Energy Information Administration on energy market impacts of alternative greenhouse gas reduction policies to support their estimates.

Finally, in responding to our first questionnaire, panelists were asked to identify key assumptions that they made in describing the estimates of the benefits and costs of their proposed actions to address climate change. In the second questionnaire, the panelists rated the importance of the assumptions in terms of affecting the benefits and costs estimates generated by integrated assessment models (see app. III for more detail). Figure 4 illustrates the opinion of panelists on the importance of various assumptions made in developing estimates using integrated assessment models. Assumptions that panelists identified as the most important were the real discount rate (interest rate used for discounting, adjusted for inflation) for assessing the benefits and costs of climate change, and the inclusion of all economic sectors in policies to address climate change. With respect to discounting future benefits and costs, we asked panelists to identify a reasonable estimate for the discount rate. Fifteen panelists responded with estimates ranging from 0 to 5 percent.

PANELISTS RATED ESTIMATES OF COSTS OF ACTIONS TO ADDRESS CLIMATE CHANGE AS MORE USEFUL THAN ESTIMATES OF BENEFITS, CITING UNCERTAINTIES ASSOCIATED WITH FUTURE IMPACTS

Citing uncertainties associated with the potential future impacts of climate change, and the difficulties of estimating their economic impacts, panelists rated cost estimates from integrated assessment models as more useful for informing congressional decision making than benefit estimates. While panelists identified challenges in estimating costs such as predicting future technological development and additional costs associated with inefficient policy designs, 10 panelists stated that the estimates of costs were quite or extremely useful, whereas the majority of panelists said that estimates of benefits were only somewhat or moderately useful. However, all of the panelists that responded to the applicable questions on this topic said that the estimates of costs and benefits from integrated assessment models were at least somewhat useful. Panelists provided a number of rationales for their opinions on integrated assessment models. For example, one panelist said that costs were easier to estimate primarily because economists have historical data that can be used to model the effect of changes in energy prices. Conversely, one panelist stated that the benefits of emissions mitigation are poorly understood because they are based on a limited understanding about the climate system. Another panelist added that even though researchers have put substantial effort into quantifying the avoided damages of actions to address climate change, they are still highly speculative.

When asked to rate the relative importance of uncertainties that may affect the estimated benefits and costs in integrated assessment models identified in the first questionnaire, on average, the panelists rated thresholds and abrupt changes in the climate system as the most important uncertainty, followed by the science of climate change, and the economic effect of actions to address climate change (see fig. 5 and app. III for more detail).[29] Additionally, the panelists identified how society will adapt to climate change as an important uncertainty affecting benefit and cost estimates. When asked how to best address risk and uncertainty in economic assessments of climate change policies, two panelists said it was important to provide additional information on uncertainties associated with low-probability, high-impact climate change events. For example, one panelist said that the limited treatment of low-probability high-impact uncertainty associated with high temperature changes is the biggest unsolved problem driving a complete economic analysis. Other panelists said it was important to provide decision makers with descriptions of the risks of climate change, including distributions of probable impact scenarios under various policy approaches. For example, one panelist said that ideally, economic assessments should fully characterize uncertainties by providing decision makers with probability distributions over prospective outcomes.

Despite the challenges associated with estimating benefits and costs using integrated assessment models, several panelists cited them as valuable tools for informing judgment on climate change policy. For example, one panelist said that integrated assessment models are the only tools that can assemble all of the factors surrounding the climate issue to assess potential emissions mitigation and adaptation strategies. When asked to identify steps that the Congress could take to help economists and other researchers address the most important uncertainties, most of the panelists that responded expressed the need for continued funding of research to improve understanding of the science and economics of climate change.

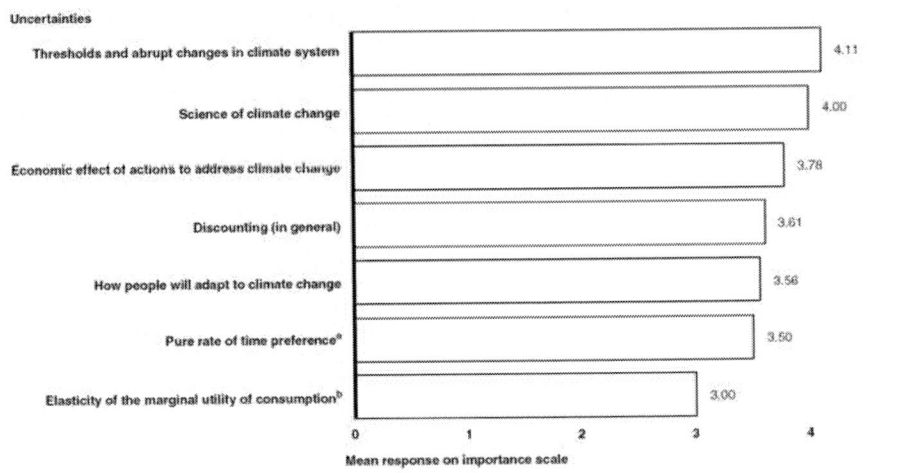

1 Not at all important
2 Somewhat important
3 Moderately important
4 Quite important
5 Extremely importan
[a] Pure rate of time preference reflects the relative weight assigned to the welfare of different generations over time.
[b] Elasticity of the marginal utility of consumption is the percentage change in welfare derived from a percentage change in consumption or income.
Source: GAO analysis.
Note: Total responses for each question may range from 15 to 18. See appendix III for additional details.

Figure 5. Mean Panelist Ratings of the Importance of Categories of Uncertainties Associated with Benefit and Cost Estimates from Integrated Assessment Models.

THE MAJORITY OF PANELISTS SAID THAT THE UNITED STATES SHOULD BEGIN TO CONTROL EMISSIONS SOON, REGARDLESS OF INTERNATIONAL PARTICIPATION

The majority of panelists said that if the United States acts unilaterally to establish a price on greenhouse gas emissions, they would not change their conclusions that their recommended actions were economically justified. While some panelists said that the United States should proceed cautiously if it acts unilaterally, 16 of 18 panelists agreed that the United States should establish a price on greenhouse gas emissions as soon as possible, regardless of the extent to which other countries adopt similar policies. Nonetheless, the majority of the panelists said that it was important for the United States to participate in international negotiations to facilitate climate agreements or to enhance the credibility or influence of the United States. In addition, some panelists noted the importance of participation by other countries, recommending that the United States act conservatively in the absence of action by other high-emitting nations. For example, one panelist said that the U.S. government should also engage in international negotiations to ensure other nations make a similar commitment. In emphasizing the importance of global participation, the panelist said that without it, a U.S. emissions reduction program would be undercut.

When asked whether U.S. action to establish a price on greenhouse gas emissions in the absence of action by other high-emitting nations (e.g., India, China, Brazil) would have a negative or positive effect on the ability of U.S. companies to compete with similar companies in other countries, 10 out of 18 panelists said the effect would be negative, 7 said it would be neither positive or negative, and 1 did not know or was not sure.

Moreover, several panelists said that energy-intensive industries such as chemicals and metals, as well as the coal sector, would experience the most negative effects if the United States established a price on greenhouse gases in the absence of such actions in other high-emitting nations.

On the other hand, one panelist remarked that the major impacts would not necessarily come from international trade, adding that the main effect of U.S. action would be on electricity production, which is not generally traded internationally. Another panelist said that competitiveness is only one factor if the United States takes action to mitigate greenhouse gas emissions and the other key countries do not. The most important factor, this panelist noted, is that in the absence of action by other high-emitters a large investment in abatement of emissions by the United States will be wasted because it will not achieve a corresponding improvement in the environment.

Chapter 4

THE PANELISTS' VIEWS ON THE STRENGTHS AND LIMITATIONS OF POLICY OPTIONS FOCUSED PRIMARILY ON THE ENVIRONMENTAL CERTAINTY OF A CAP-AND-TRADE SYSTEM VERSUS THE EFFICIENCY OF A TAX ON EMISSIONS

The panel provided their opinions on the strengths and limitations associated with various policy options to address climate change, and focused on the key trade-offs between a tax on emissions or a cap-andtrade system. Panelists first identified criteria for evaluating policy options and then discussed the strengths and limitations of different policy options within this context. The most important trade-offs identified by panelists focused on the environmental effectiveness of a cap-and-trade system versus the economic efficiency of a tax on emissions. While panelists viewed other policy options less favorably, they cited their potential as a complement to a market-based mechanism. These expert opinions should be of assistance to the Congress in weighing the potential benefits and costs of different policies for addressing climate change.

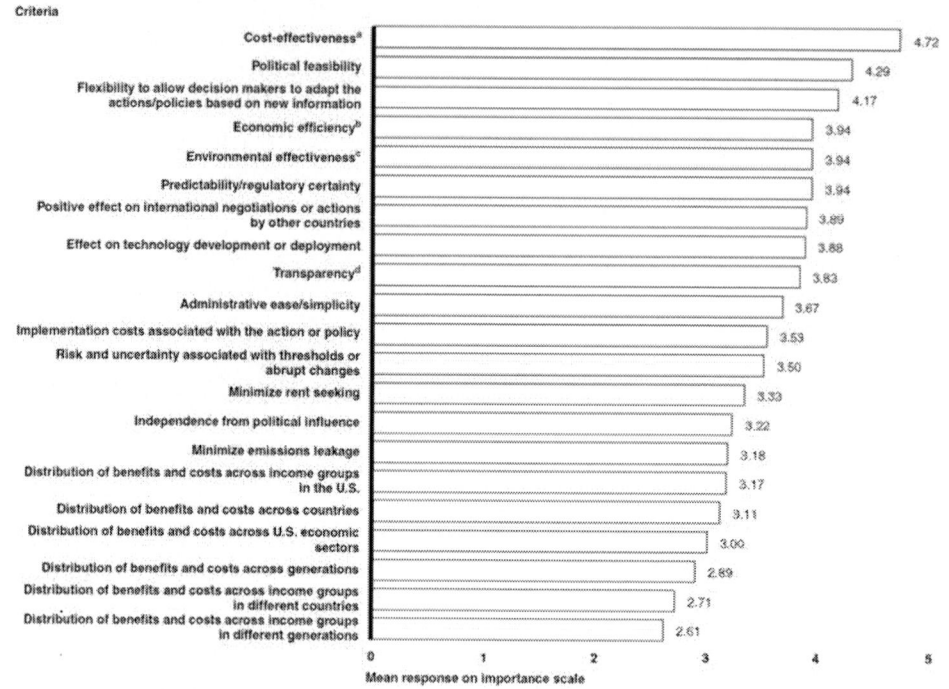

1 Not at all important
2 Somewhat important
3 Moderately important
4 Quite important
5 Extremely importan

Source: GAO analysis

Note: Total responses for each question may range from 15 to 18. See appendix III for additional details.

[a] Cost-effectiveness is the extent to which a goal is achieved in a least cost manner.

[b] Economic efficiency occurs when, from a societal perspective, production is at an efficient level (e.g. net benefits are maximized).

[c] Environmental effectiveness is the extent to which policy achieves the environmental target.

[d] Transparency is the extent to which policy's requirements and costs are visible to all parties.

Figure 6. Mean Panelist Ratings of the Importance of Criteria in Evaluating Policies for Addressing Climate Change.

PANELISTS RATED THE IMPORTANCE OF CRITERIA FOR EVALUATING POLICY OPTIONS

In our first questionnaire, panelists identified criteria that they believe the Congress should consider in evaluating the various actions and policy options for addressing climate change. In the second questionnaire, panelists rated the importance of those criteria, such as economic efficiency and environmental effectiveness. Economic efficiency is used to assess whether a policy alternative would maximize net benefits (that is, where marginal benefits equal marginal costs) to society, and environmental effectiveness means that the policy implemented has the desired environmental result. The criteria were divided into two categories: efficiency-related criteria, such as economic efficiency, environmental effectiveness, cost effectiveness, flexibility and adaptability of policies; and equity-related criteria, such as political feasibility, impact on international negotiations, and the distribution of costs and benefits among generations and countries. On average, the panelists rated cost effectiveness and political feasibility as the most important criteria (see fig. 6 and app. III for more detail).

THE MOST IMPORTANT TRADE-OFFS ARE BETWEEN THE RELATIVE EFFECTIVENESS OF CAP-AND-TRADE SYSTEMS AND THE RELATIVE EFFICIENCY OF TAXES

When asked to identify their preferred policy options, 8 of the 18 panelists said that they prefer a cap and trade with a safety valve as a way to combine some elements of both a cap-and-trade system and a tax. Some panelists noted that a traditional cap-and-trade system would provide more environmental certainty; that is, it would more likely achieve specific emissions reductions because the policy caps the total amount of emissions at a specific level. On the other hand, several panelists said a tax on emissions would be more economically efficient, citing, for example, its ability to provide certainty about marginal abatement costs associated with controlling greenhouse gases and avoiding price volatility that could occur in a permit market. In addition, some noted that a cap-and-trade program can be more administratively burdensome than a tax. Nonetheless, some of the panelists that preferred a tax said that a cap-andtrade program, especially if it included cost minimizing components, would be an acceptable second option to address climate change.

The majority of panelists supporting a safety valve cited the potential for limiting volatility as an important rationale for incorporating a safety valve into a cap-and-trade program. In addition, some panelists cited uncertainty regarding the marginal costs and marginal benefits of complying with a cap-and-trade program, and some cited the flexibility that it would provide to decision makers to respond to new information as rationales for having a safety valve. However, 3 panelists expressed concern regarding the compatibility of a safety valve with an international greenhouse gas trading system. For example, one panelist stated that under any realistic safety valve system, the permits should be good only in the country of issue. Otherwise, countries with higher market emissions prices could buy permits in the United States, undermining their respective emissions targets. Another panelist added that if the United States commits to specific emissions targets as part of an international agreement, a safety valve may affect the United States' ability to meet those commitments. If the safety valve is triggered, for example, emitters will be allowed to purchase additional permits to emit above the targeted cap for emissions, which may keep the United States from meeting agreed-to goals. Potential solutions offered by panelists included abolishing the safety valve after the emissions market has stabilized and then integrating into an international system, and incorporating a quantitatively limited safety valve where a limited number of permits were sold once the safety valve was reached.

A majority of the panelists said that the government should auction off rather than give away at least a portion of the permits under a cap-andtrade system. The means by which the government distributes allowances can have important economic implications. For example, the existing cap-and-trade programs for sulfur dioxide (a pollutant that causes acid rain) and for carbon dioxide in the European Union give away most of the allowances to regulated entities with a limited number of allowances reserved for auction. The majority of panelists believed that the government should use either a combination of auctioning and free allocations or auctioning to distribute the allowances. In addition, we asked the panelists to rate the importance of various actions to distribute the revenue generated from auctioning allowances that could offset adverse effects on consumers or particular sectors of the economy. On average, the panelists rated using the revenues to reduce the tax burden for low-income individuals as the most important way to distribute the revenue (see fig. 7 and app. III for more detail). In addition, on average, panelists rated using revenues to support research and development in various areas, such as zero- or low-carbon technologies and carbon capture and storage as at least moderately important.

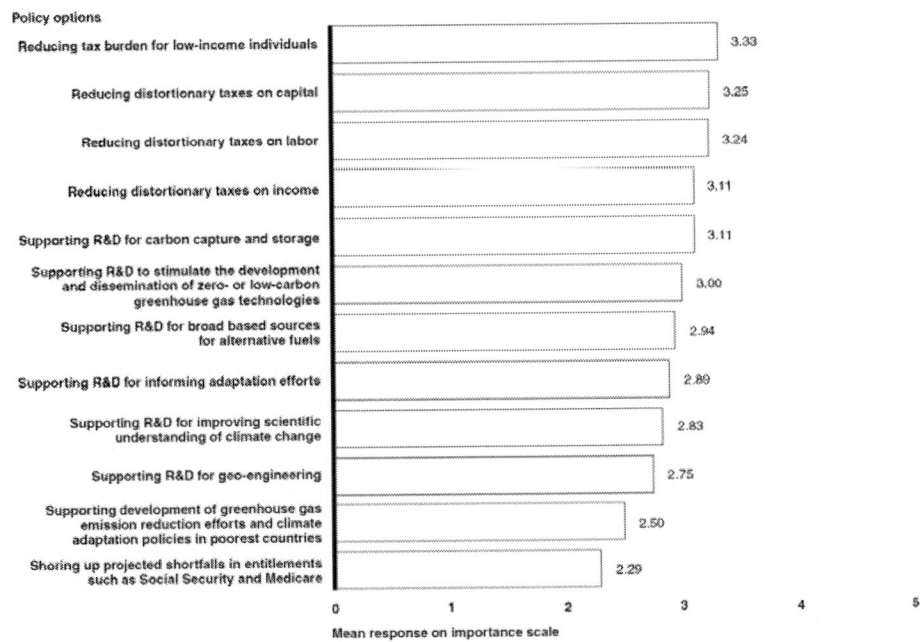

1 Not at all important
2 Somewhat important
3 Moderately important
4 Quite important
5 Extremely importan
Source: GAO analysis.
Note: Total responses for each question may range from 15 to 18. See appendix III for additional details.

Figure 7. Mean Panelist Ratings of the Importance of Ways That Revenue Generated from a Market-Based Mechanism Could Be Distributed.

THE PANELISTS VIEWED OTHER POLICY OPTIONS LESS FAVORABLY BUT CITED THEIR POTENTIAL AS A COMPLEMENT TO A MARKET-BASED MECHANISM

In the second questionnaire, we asked panelists to discuss the strengths and weaknesses of other domestic policy options that would serve to complement a tax or cap-and-trade system, but some viewed them less favorably in the absence of a market-based mechanism. In weighing the strengths and weaknesses of policy options, we asked panelists to discuss research and development of technologies, adaptation to climate change, revising efficiency standards, and reforming subsidies. For example, several panelists noted that mechanisms such as vehicle fuel efficiency standards would be unnecessary if the Congress enacted a mitigation policy to place a price on carbon. Two panelists noted that setting efficiency standards for vehicles was unlikely to yield cost-effective reductions in emissions because greater fuel efficiency lowers the cost of driving, potentially leading individuals to drive more. Table 3 reflects examples cited by the panelists of strengths and limitations of specific policy options.

Table 3. Key Strengths and Limitations of Other Policy Options Identified by Panelists

Policy option	Examples of strengths	Examples of limitations
R&D in low- or zero-carbon technologies	• May help lower long-term costs of mitigation • Promotes the development of new technologies and encourages innovation • Facilitates international technology transfer • Government-funded R&D can be disseminated quickly • Privately funded R&D may be slow to develop	• Insufficient as a policy on its own • Unnecessary in the presence of a robust mitigation policy • Uncertainty regarding feasibility, costs, andtiming of deployment • Danger of funding less useful technologies due to politics or uncertainty • Could crowd out other sources of R&D funding
Adaptation	• Essential to help federal, state, and local governments plan and prepare for unavoidable consequences • Clear role for the federal government in some sectors, such as water resources and land management	• Insufficient policy on its own • Difficult to define • Private sector has incentives to adapt without federal intervention • Federal government has little role • Will yield only moderate payoffs in terms ofreducing future costs • Removes incentive to reduce emissions or develop better energy technologies

Reforming subsidies for fuels or energy-efficient technologies	• Necessary to make current policies more efficient • Facilitates the development of better fuels and energy-efficient technologies • May result in better pricing of resources	• Insufficient policy on its own • Unnecessary in the presence of a robust mitigation policy • Inferior to mitigation for addressing climatechange • Potential for rent seeking[a] • Economically inefficient • Politically difficult
Develop or revise efficiency standards	• Could reduce the overall costs of mitigation • Important in the presence of real market failures, such as buildings and consumer appliances • Politically feasible	• Insufficient policy on its own • Unnecessary in the presence of a robust mitigation policy • Economically inefficient • Unequal burden of costs • Limited potential benefits
R&D on the basic science of climate change	• Informs adaptation needs • Guides stringency of climate policy goals • Reduces uncertainty regarding future impacts, costs, and benefits of climate change • Enhances monitoring	• Insufficient policy on its own • Unnecessary in the presence of a robust mitigation policy • Sufficient knowledge on the subject • Adequately funded at current levels • Could postpone action on mitigation • Difficult to determine the proper amount of R&D needed

Source: GAO analysis of panelists' responses.
[a] Rent seeking occurs when a party (e.g., individual or organization) seeks an economic gain (e.g., subsidy) from the government.

The panelists cited the strengths and weaknesses of public investments in the development and deployment of technologies that could reduce emissions. Some suggested such efforts could complement a market-based mitigation policy by, for example, lowering the costs of controlling emissions and enhancing the likelihood of achieving long-term emissions targets, while others noted that public investment may be unnecessary with a mitigation policy in place. In addition, some panelists said that while they supported research and development investment, it was not enough on its own without additional policies to address climate change. Further, several panelists also expressed concern about the government picking what it perceives as "winning" technologies rather than funding a broader research and development program.

A majority of panelists said that funding efforts to adapt to the impacts of climate change were at least moderately important. Some panelists identified actions the federal government could take to prepare for climate change impacts such as reforming insurance subsidy programs in areas vulnerable to natural disasters like hurricanes or flooding and creating capacity within federal agencies to protect against and react to climate change impacts. When discussing the strengths and weaknesses of actions to adapt to climate change, several panelists noted that implementing adaptation policies may help reduce vulnerabilities when faced with inevitable climate change or in the event of possible catastrophic climate change. One panelist referred to adaptation as a risk management strategy, or form of insurance, against uncertain future events resulting from climate change. Some panelists noted that a clear federal role exists for certain sectors, such as water resource management and property rights protection, which could require additional resources for infrastructure development, research, and adapting practices to use alternative methods for distributing water and managing federal lands. Another panelist said that implementing adaptation policies could also help ensure greater international and economic equity, since some areas likely to be affected by climate change are underdeveloped, economically disadvantaged, or vulnerable in some way to impacts. However, some panelists stated that adaptation was difficult to define, making the federal role unclear, and others said that incentives for adaptation may already exist, limiting the need for a federally directed adaptation policy.

Panelists identified the strengths and limitations of other policy options to address climate change, such as subsidy reform for alternative fuels or energy-efficient technologies and research and development into the basic science of climate change. For example, in rating the importance of additional actions for the Congress to consider, 12 panelists rated reforming subsidies for alternative fuels or energy-efficient technologies as at least moderately important, and several panelists noted that subsidies for alternative fuels or energy-efficient technologies are economically inefficient or insufficient on their own without an appropriate mitigation policy. Regarding research and development on the basic science of climate change, several panelists said that it could help inform adaptation efforts and reduce uncertainty of costs or benefits by improving understanding of the potential dangers from climate change.[30] However, some panelists noted that additional funding of research and development on the basic science of climate change is not warranted, citing reasons such as a belief that it is adequately funded at current levels, it could be used as an excuse to delay taking action to mitigate climate change, and that the priority should instead be placed on adaptation efforts and monitoring.

John B. Stephenson
Director,
Natural Resources and Environment

APPENDIX I: SCOPE AND METHODOLOGY

To address the first and second objectives, we (1) reviewed relevant climate change academic literature and documents developed by federal agencies, (2) met with agency officials from the Environmental Protection Agency (EPA), the Department of Energy (DOE) including the Energy Information Administration (EIA), the Department of Commerce including the National Oceanic and Atmospheric Administration (NOAA), the United States Department of Agriculture (USDA), the National Aeronautic and Space Administration (NASA), and the Council of Economic Advisors (CEA), and (3) obtained expert opinion on the economic effects of actions to address climate change and the strengths and weaknesses of those actions using a virtual panel on the Internet.

To structure and gather expert opinions from the panel, we employed a modified version of the Delphi method. The method is based on a structured process for collecting and distilling knowledge from a group of experts by means of a series of questionnaires.[1] Used to support informed decision making, the Delphi method was first developed at the RAND Corporation in the 1950s. One of the strengths of this approach is its flexibility, and while first used in a live group discussion format, the method is easily modified for various settings. The modified process we employed utilized two Web-based questionnaires, and incorporated an iterative and controlled feedback process to gather the experts' opinions. Specifically, experts' responses to the first questionnaire were used to create the questions for the second, allowing the experts to consider the opinions and issues raised by other panelists when responding to the second round of questions. Also, by using a Web-based process, we were able to overcome some of the potential biases associated with group discussions. These biasing effects

---

[1] Adler, Michael, and Eric Ziglio, eds. *Gazing into the Oracle: The Delphi Method and Its Application to Social Policy and Public Health* (Bristol, Pennsylvania: 1996).

include the potential dominance of individuals and group pressure for conformity. Moreover, by creating a virtual panel, we were able to include more experts than would have been possible with a live panel. While the method has these strengths, there are some potential limitations. For example, there is considerable reliance on the active participation of the panelists, which can vary widely, and some panelists may not complete the entire study. In addition, the results of the iterative process are limited to the issues, topics, and responses generated by those participating; thus some topics or viewpoints may not be considered in the process. To mitigate the latter limitation, we added critical questions to the second questionnaire that were not discussed in the first round based on our review of the literature.

We contracted with the National Academy of Sciences (NAS) to select and recruit a panel of experts with a range of in-depth experience assessing the economic impacts of climate change policy. Participants were to (1) have expertise in modeling and analyzing benefits, costs, and uncertainties using integrated assessment methods; (2) have knowledge of policies for mitigating climate change; (3) have knowledge of the economic trade-offs associated with different policies for mitigating climate change; and (4) be affiliated with U.S.-based institutions including academia, the federal government, and other research-oriented entities. To select the experts, we provided NAS with a preliminary list of potential panelists that we identified in our review of the literature. Taking these names into account, NAS developed a list of 37 who met our criteria. We reviewed and agreed with the list of names, and NAS sent 25 individuals an electronic letter via e-mail inviting them to participate in the study along with a description of the project. Invitees were given the option of discussing the project further with a project representative before deciding whether or not to participate, which some chose to do. Of the 25 panelists NAS recruited to participate, 21 agreed and were sent the first questionnaire. Nineteen responded to the first questionnaire, and 18 responded to the second.[2] All of the experts who participated completed a form stating that they had no conflicts of interest that would compromise their ability to participate in the panel.

Prior to the posting of the questionnaires, we conducted a series of pretests with internal and external experts, including two panel participants. The goals of the pretests were to check that (1) the questions were clear and unambiguous and (2) terminology was used correctly. We made changes to the content and format

---

[2] The responses presented in this report are from the second questionnaire, supplemented with examples and anecdotes from the open-ended responses in the first round, and represent the views of the 18 panelists who participated in both phases of the panel.

of both questionnaires as necessary during the pretesting processes. We also conducted usability tests of both questionnaires for the Internet to ensure operability. For each phase of the Delphi, we posted a questionnaire on the Internet. Panel members were notified of the availability of the questionnaire with an e-mail message. The e-mail message contained a unique user name and password that allowed each respondent to log on and fill out a questionnaire but did not allow respondents access to the questionnaires of others.

In the first phase, we asked panelists to provide responses to three open-ended questions on the economics of actions to address climate change developed from an extensive literature review. We asked the panelists to (1) identify what actions, if any, the Congress should consider to address climate change; (2) provide estimates of the benefits and costs of their recommended actions; and (3) identify criteria that are appropriate for evaluating the potential actions. In addition, we asked the panelists to provide citations to support their responses.

After the first questionnaire was completed, we performed a content analysis of the open-ended responses. Using the themes and topics that were discussed, as well as the varying opinions of the panel, the second questionnaire was constructed. For example, panelists were asked to rate the importance of the various assumptions and uncertainties associated with integrated assessment models that the experts collectively identified in the first questionnaire. Using this approach, panelists could provide their opinions on the topics that others had raised, and areas of agreement and disagreement could be identified. In addition, this approach allowed for the panelists to reevaluate their original responses in light of the responses of the whole group. While these first round responses were the primary source for developing the second questionnaire, literature cited in support of their responses, as well as information gathered from GAO's literature review were also incorporated as necessary. The second phase questionnaire included mostly closed-ended questions, with a limited number of open-ended questions, and encapsulated panelists' views on preferred policy options, potential benefits and costs, key uncertainties, and the strengths and weaknesses of different policy options.

Panel members had approximately 4 weeks between July and August of 2007 to complete their questionnaires in the first phase of the panel, approximately 4 weeks in October and November of 2007 to complete their questionnaires in the second phase of the panel, and approximately 4 weeks to provide comment (at their discretion) on a summary of their second round responses in January of 2008. Selected questions and aggregated responses from the second phase are presented in appendix III. In addition, we asked several follow-up questions requesting that panelists clarify their responses or elaborate on critical policy issues. While we display only the quantitative, closed-ended responses, we also relied on the responses to the qualitative, open-ended questions to inform our findings in this book. GAO provided a summary of the findings of this book and briefed representatives from the CEA, the Council on Environmental Quality (CEQ), EPA, and DOE on the results of the panel prior to issuing this book. The views expressed by the panel members do not necessarily represent the views of GAO.

APPENDIX II: SELECTED CHARACTERISTICS OF PANELISTS' PREFERRED POLICY OPTIONS FOR ADDRESSING CLIMATE CHANGE

Listed below are selected panelist policy recommendations for implementing a market-based mechanism to place a price on greenhouse gas emissions.

Table 4. Selected Characteristics of Panelists' Preferred Policy Options for Addressing Climate Change

Panelist	Preferred market-based policy	Greenhouse gases covered	Date of implementation	Scope	Point of regulation[a]	Initial price range[b]	Safety valve price[c]	Complementary policy actions
Panelist 1	Cap and trade	All gases	2010-2015	Economywide	Combination	> 14	N/A	(1) R&D on low carbon energy technologies; (2) energy efficiency and renewable portfolio standards
Panelist 2	Cap and trade	All gases	2010-2015	Economywide	Upstream	0.3-3	N/A	Not specified
Panelist 3	Cap and trade	All gases	2010-2015	Economywide	Combination	21-30	N/A	(1) R&D subsidies; (2) adaptation

Table 4. Continued

Panelist 4	Cap and trade w/ safety valve	Carbon dioxide	2010-2015	Economywide	Upstream	21-30	Don't know	Not specified
Panelist 5	Cap and trade w/ safety valve	All gases	Before 2010	Economywide	Upstream	0.3-3	0.55	(1) International negotiation and assistance
Panelist 6	Cap and trade w/ safety valve	All gases	2010-2015	Economywide	Downstream	3-5	3.81	(1) Do away with import tariffs and agricultural subsidies;(2) tax breaks for private sector research and development; (3) remove perverse subsidies; (4) declassify energy saving and lightweight technologies for export to China and India
Panelist 7	Cap and trade w/ safety valve	Carbon dioxide	Before 2010	Economywide	Combination	3-5	5	(1) Funding R&D; (2) funding adaptation; (3) reform policies that discourage energy efficiency
Panelist 8	Cap and trade w/ safety valve	All gases	2010-2015	Economywide	Upstream	11-20	3 to 4 times the initial price of emissions	(1) Negotiate international agreements with major greenhouse gas emitters; (2) invest in pilot demonstration projects for carbon capture and storage and renewable fuels.

Appendix II

Panelist 9	Cap and trade w/ safety valve	All gases	2010-2015	Economywide	Upstream	11-20	12 to 20	(1) Support for technology research, development, and deployment; (2) support for developing countries; (3) international engagement to leverage our action for action by other countries.
Panelist 10	Cap and trade w/ safety valve	All gases	Before 2010	Economywide including land use	Combination	1-10	2.70	(1) Limit loopholes that could arise in policies that focus on specific sectors or groups of economic actors.
Panelist 11	Cap and trade w/ safety valve	All gases	Before 2010	Economywide	Downstream	11-20	25	(1) Research and development; (2) technology transfer (both emphasizing adaptation and emissions reduction strategies).
Panelist 12	Tax	Carbon dioxide	Before 2010	Economywide	Upstream	11-20	N/A	(1) Support carbon capture and storage; (2) sponsor research and development in alternative energy sources; (3) sponsor adaptation at home and abroad.
Panelist 13	Tax	All gases	Before 2010	Economywide	Downstream	0.3-3	N/A	Not specified
Panelist 14	Tax	All gases	Before 2010	Economywide	Combination	6-8	N/A	(1) Basic research in the underlying science for low-carbon energy
Panelist 15	Tax	Carbon dioxide first, then expand to all	2010-2015	Economywide	Upstream	11-20	N/A	(1) Adaptation; (2) R&D (including geo-engineering)

Table 4. Continued

Panelist 16	Tax	All gases	2010-2015	Economywide	Mostly up-stream, midstream for utilities and carbon capture and storage incorporating carbon sinks from land use	11-20	N/A	(1) Independently directed effort to support R&D; (2) international assistance and negotiations to facilitate action by other nations
Panelist 17	Tax	No answer	Before 2010	Economywide	Downstream	41-50	N/A	Not specified
Panelist 18	Tax	All gases	Before 2010	Economywide	Upstream	N/A[d]	N/A	Not specified

Source: GAO.

Note: N/A stands for not applicable.

[a] Point of regulation in the economy. Combination means the policy would have both upstream and downstream provisions.

[b] Initial price range per metric ton of carbon dioxide equivalent in 2007 dollars. [c] Safety valve price per metric ton of carbon dioxide equivalent in 2007 dollars.

[d] One panelist preferred a policy that set a price for each individual greenhouse gas as opposed to one price for all emissions.

APPENDIX III: SELECTED QUESTIONS AND EXPERT RESPONSES

Listed below are questions and summary panelist responses supporting figures 2 through 7 of this book. Panelists were asked to rate the importance of various items related to policy approaches and rationales for addressing climate change (Mean responses were calculated by assigning the following values to the rated level of importance: 1 = not at all, 2 = somewhat, 3 = moderately, 4 = quite, and 5 = extremely).

Question 1: In addition to establishing a price on anthropogenic sources of emissions using either a tax or cap and trade system (or other regulatory approach), panelists identified the following list of actions that Congress might also consider as part of a broader portfolio of actions. How important is it that Congress consider each action to address climate change?

Additional actions to address climate change	Panelist responses							
	Rated level of importance							
	Not at all	Somewhat	Moderately	Quite	Extremely	Don't know / no response	Number of responses	Mean level of importance
Participating in international negotiations to facilitate participation in climate agreements or to enhance U.S. credibility/influence	0	1	2	2	12	1	17	4.47

Table. Continued

Funding R&D and/or deployment of low- or zero-carbon technologies	1	5	2	5	5	0	18	3.44
Creating an independent oversight board to establish, monitor, and revise the mechanism for controlling greenhouse gases (tax or cap and trade)	3	3	1	4	5	2	16	3.31
Providing international assistance to developing countries	1	5	5	3	4	0	18	3.22
Reforming subsidies for alternative fuels or energy-efficient technologies	1	4	5	5	2	1	17	3.18
Funding efforts to adapt to impacts of climate change by reforming agricultural subsidy programs, land use practices, flood control, etc.	1	5	4	6	2	0	18	3.17
Developing or revising technology efficiency standards (for fuels or energy use)	6	1	3	6	1	1	17	2.71
Funding private sector R&D to achieve cost-effective emissions reductions in developing countries	4	7	3	3	1	0	18	2.44
Funding R&D on domestic and international public health impacts that could arise as a result of climate change	2	12	1	2	1	0	18	2.33
Levying a graduated tax on cars and light trucks (including sport utility vehicles), based on fuel economy	7	3	5	2	1	0	18	2.28
Minimize emissions leakage	1	3	6	6	1	1	17	3.18

Criteria	Not at all	Somewhat	Moderately	Quite	Extremely	Don't know / no response	Number of responses	Mean level of importance
Distribution of benefits and costs across income groups in the United States	0	4	8	5	1	0	18	3.17
Distribution of benefits and costs across countries	0	3	11	3	1	0	18	3.11
Distribution of benefits and costs across U.S. economic sectors	0	7	5	5	1	0	18	3.00
Distribution of benefits and costs across generations	1	7	5	3	2	0	18	2.89
Distribution of benefits and costs across income groups in different countries	0	9	4	4	0	1	17	2.71
Distribution of benefits and costs across income groups in different generations	1	10	2	5	0	0	18	2.61

Source: GAO.

Question 2: In the first round of questions, panelists identified criteria that the Congress should consider in evaluating the various actions and policy options for addressing climate change. The list below summarizes these criteria. How important would you say that the following criteria are in evaluating the various policies for addressing climate change?

	Panelist responses							
	Rated level of importance							
Criteria	Not at all	Somewhat	Moderately	Quite	Extremely	Don't know / no response	Number of responses	Mean level of importance
Cost-effectiveness	0	0	0	5	13	0	18	4.72
Political feasibility	0	1	1	7	8	1	17	4.29
Flexibility to allow decision makers to adapt the actions/policies based on new information	0	0	4	7	7	0	18	4.17
Economic efficiency	0	1	5	6	6	0	18	3.94

Table. Continued

Environmental effectiveness	0	2	3	7	6	0	18	3.94
Predictability/regulatory certainty information	0	2	4	5	7	0	18	3.94
Positive effect on international negotiations or actions by other countries	0	3	1	9	5	0	18	3.89
Effect on technology development or deployment	0	1	4	8	4	1	17	3.88
Transparency	0	2	3	9	4	0	18	3.83
Administrative ease/simplicity	0	2	4	10	2	0	18	3.67
Implementation costs associated with the action or policy	0	3	4	8	2	1	17	3.53
Risk and uncertainty associated with thresholds or abrupt changes	1	3	3	8	3	0	18	3.50
Minimize rent seeking	2	2	6	4	4	0	18	3.33
Independence from political influence	1	4	6	4	3	0	18	3.22
Minimize emissions leakage	1	3	6	6	1	1	17	3.18
Distribution of benefits and costs across income groups in the United States	0	4	8	5	1	0	18	3.17
Distribution of benefits and costs across countries	0	3	11	3	1	0	18	3.11
Distribution of benefits and costs across U.S. economic sectors	0	7	5	5	1	0	18	3.00
Distribution of benefits and costs across	1	7	5	3	2	0	18	2.89

generations								
Distribution of benefits and costs across income groups in different countries	0	9	4	4	0	1	17	2.71
Distribution of benefits and costs across income groups in different generations	1	10	2	5	0	0	18	2.61

Source: GAO.

Question 3: Panelists identified several actions that Congress should consider in deciding how to distribute the revenue that would be generated from a federal program to establish a price on greenhouse gases (from an imposition of a tax or from an auction of permits). How important is it for Congress to distribute the generated revenue to each of these categories?

	Panelist responses							
	Rated level of importance							
Uses for revenue distribution	Not at all	Somewhat	Moderately	Quite	Extremely	Don't know / no response	Number of responses	Mean level of importance
Reducing tax burden for low-income individuals	1	4	5	4	4	0	18	3.33
Reducing distortionary taxes on capital	1	1	8	5	1	2	16	3.25
Reducing distortionary taxes on labor	0	3	9	3	2	1	17	3.24
Reducing distortionary taxes on income	0	4	9	4	1	0	18	3.11
Supporting R&D for carbon capture and storage	2	4	6	2	4	0	18	3.11
Supporting R&D to stimulate the development and dissemination of zero- or low-carbon/greenhouse gas technologies	2	4	5	6	1	0	18	3.00

Source: GAO.

Table. Continued

Supporting R&D for broad-based sources for alternative fuels	3	3	7	2	3	0	18	2.94
Supporting R&D for informing adaptation efforts	3	5	4	3	3	0	18	2.89
Supporting R&D for improving scientific understanding of climate change	2	6	5	3	2	0	18	2.83
Supporting R&D for geo-engineering	3	4	5	2	2	2	16	2.75
Supporting development of greenhouse gas emission reduction efforts and climate adaptation policies in poorest countries	2	9	4	2	1	0	18	2.50
Shoring up projected shortfalls in entitlements such as Social Security and Medicare	6	5	2	3	1	1	17	2.29

Source: GAO.

Question 4: Many panelists said that it is difficult to estimate the potential benefits associated with emissions mitigation partly because of uncertainty about the impact of climate change and the actions that could be used to address it. In some cases, the panelists identified general categories of benefits that could accrue from various policy options. These categories are below. How important are the following categories of potential benefits as a rationale for addressing climate change?

	Panelist responses							
	Rated level of importance							
Categories of benefits of actions to address climate change	Not at all	Somewhat	Moderately	Quite	Extremely	Don't know / no	Number of responses	Mean level of importance
Avoided climate change damages (including flooding, extreme weather events, crop damage, impacts on sensitive ecosystems, public health, and species loss)	0	0	1	11	5	1	17	4.24

Key assumptions	Not at all	Somewhat	Moderately	Quite	Extremely	Don't know / no response	Number of responses	Mean level of importance
Reducing risk or the likelihood of extreme events associated with climate change	1	0	2	8	6	1	17	4.06
Establishing a price signal to influence market or individual behavior	3	1	0	2	10	2	16	3.94
Facilitation of international cooperation on climate change	3	1	0	5	7	2	16	3.75
Development of a regulatory system for greenhouse gas emissions	3	2	2	4	5	2	16	3.38
Inducing investment in developing countries	4	3	4	1	3	3	15	2.73
Development, deployment, and, more generally, creating incentives for businesses and consumers to invest in low- or zero-carbon technologies	5	3	1	6	1	2	16	2.69
Reducing risk of international conflict over natural resources and territory	4	2	7	2	1	2	16	2.63

Question 5: Panelists identified several key assumptions that they made in estimating the benefits and costs using integrated assessment models. How important are the following assumptions in terms of affecting the estimated benefits and costs in integrated assessment models?

	Panelist responses							
	Rated level of importance							
Key assumptions	Not at all	Somewhat	Moderately	Quite	Extremely	Don't know / no response	Number of responses	Mean level of importance
Real discount rate for assessing benefits and costs of addressing climate change	0	1	2	3	11	1	17	4.41

Inclusion of all economic sectors	0	1	1	9	6	1	17	4.18
Baseline economic growth	0	0	2	11	3	2	16	4.06
Degree of international participation	0	2	1	8	6	1	17	4.06
Climate sensitivity	0	0	3	10	4	1	17	4.06
Baseline energy use	0	0	3	11	3	1	17	4.00
Rate of technological change	0	1	3	9	3	2	16	3.88
Likelihood of abrupt change or extreme events	0	2	4	7	4	1	17	3.76
Energy price projections	0	2	5	8	2	1	17	3.59
Price elasticities	0	0	7	9	0	2	16	3.56
Coverage across greenhouse gases	0	2	7	5	3	1	17	3.53
Baseline population growth	0	1	10	5	1	1	17	3.35
Baseline adaptation	1	4	4	6	2	1	17	3.24
Efficient use of revenue by reducing distortionary taxes	0	5	7	4	1	1	17	3.06
Ancillary public health benefits	1	3	12	1	0	1	17	2.76

Source: GAO.

Question 6: Numerous respondents identified the importance of considering risk and uncertainty in identifying policy options for the Congress to consider. Below is a list of the key uncertainties that panelists identified. In general, how important are each of these uncertainties in terms of affecting the estimated benefits and costs in integrated assessment models?

Appendix III

Key uncertainties	Panelist responses							
	Rated level of importance					Don't know / no response	Number of responses	Mean level of importance
	Not at all	Somewhat	Moderately	Quite	Extremely			
Thresholds and abrupt changes in climate system	0	1	4	5	8	0	18	4.11
Science of climate change	0	0	5	8	5	0	18	4.00
Economic effect of actions to address climate change	0	2	4	8	4	0	18	3.78
Discounting (in general)	1	4	2	5	6	0	18	3.61
How people will adapt to climate change	0	2	6	8	2	0	18	3.56
Pure rate of time preference	1	4	3	5	5	0	18	3.50
Elasticity of the marginal utility of consumption	2	4	4	6	1	1	17	3.00

Source: GAO.

APPENDIX IV: PANEL OF EXPERTS

Joseph Aldy, Resources for the Future
James Edmonds, Pacific Northwest National Laboratory
Richard Howarth, Dartmouth College
Bruce McCarl, Texas A&M University
Robert Mendelsohn, Yale University
William Nordhaus, Yale University
Sergey Paltsev, Massachusetts Institute of Technology
William Pizer, Resources for the Future
David Popp, Syracuse University
John Reilly, Massachusetts Institute of Technology
Roger Sedjo, Resources for the Future
Kathleen Segerson, University of Connecticut
Brent Sohngen, Ohio State University
Robert Stavins, Harvard University
Richard Tol, Economic and Social Research Institute
Martin Weitzman, Harvard University
Peter Wilcoxen, Syracuse University
Gary Yohe, Wesleyan University

APPENDIX V: BIBLIOGRAPHY OF SELECTED LITERATURE REVIEWED BY GAO

Asterisks identify citations provided by panelists in their responses.

Aldy, Joseph E., Scott Barrett, and Robert N. Stavins. "Thirteen Plus One: A Comparison of Global Climate Policy Architectures." *Climate Policy*, vol. 3 (2003): 373-397.

*Ahmed, Rasha, and Kathleen Segerson. *Emissions Control and the Regulation of Product Markets: The Case of Automobiles*. Working Paper, Department of Economics, University of Connecticut: July 2007.

*Antle, J. M., S. M. Capalbo, K. Paustian and M. K. Ali. "Estimating the Economic Potential for Agricultural Soil Carbon Sequestration in the Central United States Using an Aggregate Econometric-Process Simulation Model," *Climatic Change*, vol. 80 (2007): 145-171.

Ashenfelter, Orley, and Karl Storchmann. *Using a Hedonic Model of Solar Radiation to Assess the Economic Effect of Climate Change: The Case of Mosel Valley Vineyards*. Working Paper 12380, National Bureau of Economic Research. Cambridge: July 2006.

Azar, Christian, and Kristian Lindgren. "Catastrophic Events and Stochastic Cost-Benefit Analysis of Climate Change: Editorial." *Climatic Change*, vol. 56, (2003): 245-255.

*Brown, P. G. "Toward an Economics of Stewardship: The Case of Climate." *Ecological Economics*, Vol. 26 (1998): 11-21.

*Burtraw et al. "Ancillary Benefits of Reduced Air Pollution in the United States from Moderate Greenhouse Gas Mitigation Policies in the Electricity Sector." *Journal of Environmental Economics and Management*, 45 (2003): 650-73.

*United States Climate Change Science Program. *Scenarios of Greenhouse Gas Emissions and Atmospheric Concentrations*. 2007.

*Cline, William R. *The Economics of Global Warming*. Washington, D.C.: Institute for International Economics, 1992.

Congressional Budget Office. *Evaluating the Role of Prices and R&D in Reducing Carbon Dioxide Emissions*. Pub. 2731. Washington, D.C.: September 2006.

Congressional Budget Office. *Trade-offs in Allocating Allowances for Carbon Dioxide Emissions*. Washington, D.C.: April 25, 2007.

Congressional Budget Office. *Uncertainty in Analyzing Climate Change: Policy Implications*. Washington, D.C.: January 2005.

Congressional Research Service. *Climate Change: Design Approaches for a Greenhouse Gas Reduction Program*. RL33799. Washington, D.C.: Updated January 16, 2007.

Congressional Research Service. *U.S. Global Climate Change Policy: Evolving Views on Cost, Competitiveness, and Comprehensiveness*. RL30024. Washington, D.C.: Updated January 29, 2007.

DeCanio, Stephen J., Richard B. Howarth, Alan H. Sanstad, Stephen H. Schneider, and Starley L. Thompson. *New Directions in the Economics and Integrated Assessment of Global Climate Change*. Arlington, Va.: Pew Center on Global Climate Change: October 2000.

Edmonds, Jae, John Clarke, James Dooley, Son H. Kim, and Steven J. Smith, "Stabilization of Carbon Dioxide in a B2 world: Insights on the Roles of Carbon Capture and Disposal, Hydrogen, and Transportation Technologies." *Energy Economics*, vol. 26, (July 2004): 517-537.

Edmonds, Jae, Marshall Wise, and David W. Barns. "Carbon Coalitions: The Cost and Effectiveness of Energy Agreements to Alter Trajectories of Atmospheric Carbon Dioxide Emissions." *Energy Policy*, vol. 23, no. 4/5 (1995): 309-335.

Energy Information Administration. *Energy Market and Economic Impacts of a Proposal to Reduce Greenhouse Gas Intensity with a Cap and Trade System*. SR/OIAF/2007-01. Washington, D.C.: January, 2007.

Environmental Protection Agency. *Tools of the Trade: A Guide to Designing and Operating a Cap and Trade Program for Pollution Control*. EPA430-B-03-002. Washington, D.C.: June, 2003.

Fankhauser, Samuel. *Valuing Climate Change: The Economics of the Greenhouse*. London, UK: Earthscan Publications Limited, 1995.

Fankhauser, Samuel, and Richard S. J. Tol. "On Climate Change and Economic Growth." *Resource and Energy Economics*, vol. 27, (2005): 1-17.

*Fischer, Carolyn, and Richard D. Morgenstern. "Carbon Abatement Costs: Why the Wide Range of Estimates?" *The Energy Journal*, vol. 27, no. 2 (2006):73-86.

*Fischer, Carolyn, and Richard Newell. "Environmental and Technology Policies for Climate Mitigation." *Resources for the Future Discussion Paper*, RFF DP 04-05, February 2007.

Gerlagh, Reyer, and Bob van der Zwaan. "Options and Instruments for a Deep Cut in Carbon Dioxide Emissions: Carbon Dioxide Capture or Renewables, Taxes or Subsidies?" *The Energy Journal*, vol. 27, no. 3 (2006).

Gurgel, Angelo C., Sergey Paltsev, John M. Reilly, and Gilbert E. Metcalf. *U.S. Greenhouse Gas Cap-and Trade Proposals: Application of a Forward-Looking Computable General Equilibrium Model*. Report No. 150. Cambridge, Mass.: MIT Joint Program on the Science and Policy of Global Change, June 2007.

Heal, Geoffrey, and Bengt Kristrom. "Uncertainty and Climate Change." *Environmental and Resource Economics*, vol. 22 (2002): 3-39.

Helm, Dieter. "The Assessment: Climate Change Policy." *Oxford Review of Economic Policy*, vol. 19, no. 3 (November 2003):349-361.

Hope, Chris. "The Marginal Impact of CO2 from PAGE2002: An Integrated Assessment Model Incorporating the IPCC's Five Reasons for Concern." *The Integrated Assessment Journal*, vol. 6, iss. 1 (2006): 19-56.

*Howarth, R. B. "Against High Discount Rates." In *Perspectives on Climate Change Science, Economics, Politics, Ethics*, edited by W. Sinnott-Armstrong and R. B. Howarth. Amsterdam: Elsevier, 2005.

Intergovernmental Panel on Climate Change. *Climate Change 1995: Economic and Social Dimensions of Climate Change*. Contribution of Working Group III to the Second Assessment Report of the Intergovernmental Panel on Climate Change. Cambridge, UK: Cambridge University Press, 1995.

*Intergovernmental Panel on Climate Change. "Summary for Policymakers," in *Climate Change 2001: The Scientific Basis*. Cambridge, UK: Cambridge University Press, 2001.

*Intergovernmental Panel on Climate Change. "Summary for Policymakers," in *Climate Change 2001: Impacts, Adaptation, and Vulnerability*. Cambridge, UK: Cambridge University Press, 2001.

*Intergovernmental Panel on Climate Change. "Summary for Policymakers," in *Climate Change 2007: The Physical Science Basis*. Cambridge, UK: Cambridge University Press, 2007.

Intergovernmental Panel on Climate Change. *Climate change 2007: Climate Change Impacts, Adaptation and Vulnerability.* Cambridge, UK: Cambridge University Press, 2007.

*Intergovernmental Panel on Climate Change. *Climate change 2007: Mitigation of Climate Change.* Cambridge, UK: Cambridge University Press, 2007.

*Interlaboratory Working Group on Energy-Efficient and Clean Energy Technologies. *Scenarios for a Clean Energy Future.* Oak Ridge, Tenn., and Berkeley, Calif.: Oak Ridge National Laboratory and Lawrence Berkeley National Laboratory, 2000.

*Jaffe, Adam B., Richard G. Newell, and Robert N. Stavins, "Technology Policy for Energy and the Environment." In *Innovation Policy and the Economy*, vol 4, edited by Adam B. Jaffe, Josh Lerner, and Scott Stern. MIT Press: Cambridge, Mass., pp. 35-68, 2004.

*Jones, R., and G.W. Yohe (2007), "Applying Risk-Analytic Techniques to the Integrated Assessment of Climate Policy Benefits," Produced for the Global Forum on Sustainable Development on the Economic Benefits of Climate Change Policies, July 2006, Paris, France.

Jorgenson, Dale W., Richard J. Goettle, Brian H. Hurd, and Joel B. Smith. *U.S. Market Consequences of Global Climate Change.* Arlington, Va.: Pew Center on Global Climate Change, 2004.

*Kammen, Daniel M., and Gregory F. Nemet. "Reversing the Incredible Shrinking Energy R&D Budget." *Issues in Science and Technology*, 22 (2005): 84-88.

Kavuncu, Y. Okan, and Shawn D. Knabb. "Stabilizing Greenhouse Gas Emissions: Assessing the Intergenerational Costs and Benefits of the Kyoto Protocol." *Energy Economics*, vol. 27 (2005):369-386.

Keller, Klaus, Benjamin M. Bolker, and David F. Bradford. "Uncertain Climate Thresholds and Optimal Economic Growth." *Journal of Environmental Economics and Management*, vol. 48 (2004): 723-74 1.

Keller, Klaus, Matt Hall, Seung-Rae Kim, David F. Bradford, and M. Oppenheimer. "Avoiding Dangerous Anthropogenic Interference with the Climate System." *Climatic Change*, vol. 73 (2005): 227-238.

Kelly, David L., and Charles D. Kolstad. "Integrated Assessment Models for Climate Change Control." In *International Yearbook of Environmental and Resource Economics 1 999/2000: A Survey of Current Issues*, edited by Henk Folmer and Tom Tietenberg, pages 171-197. Cheltenham, UK: Edward Elgar, 1999.

*Knetsch, J. L., "Gains, Losses, and the US-EPA Economic Analyses Guidelines: A Hazardous Product?" *Environmental and Resource Economics* 32 (2005): 91-1 12.

Kolstad, Charles, and Michael Toman. *The Economics of Climate Policy.* Discussion Paper 00-40REV. Washington, D.C.: Resources for the Future, June 2001.

Lange, Andreas, Carsten Vogt, and Andreas Ziegler. "On the Importance of Equity in International Climate Policy: An Empirical Analysis." *Energy Economics*, vol. 29 (2007): 545-562.

Lewandrowski, Jan, Mark Peters, Carol Jones, Robert House, Mark Sperow, Marlen Eve, and Keith Paustian. *Economics of Sequestering Carbon in the U.S. Agricultural Sector.* Technical Bulletin TB 1909. Washington, D.C.: U.S. Department of Agriculture Economic Research Service, March, 2004.

Lind, Robert C. "Intergenerational Equity, Discounting, and the Role of Cost-Benefit Analysis in Evaluating Global Climate Policy." *Energy Policy*, vol. 23, no. 4/5 (1995): 379-389.

Maddison, David. "A Cost-Benefit Analysis of Slowing Climate Change." *Energy Policy*, vol. 23, no. 4/5 (1995): 337-346.

Manne, Alan, Robert Mendelsohn, and Richard Richels. "MERGE : A Model for Evaluating Regional and Global Effects of GHG Reduction Policies." *Energy Policy*, vol. 23, (1995): 17-34.

Manne, Alan, and Richard Richels, "The Impact of Learning-by-Doing on the Timing and Costs of CO_2 Abatement." *Energy Economics*, vol. 26, issue 4 (2004): 603-619.

*Manne, Alan, and Richard Richels. "The Kyoto Protocol: A Cost-Effective Strategy for Meeting Environmental Objectives?" *The Energy Journal* 20 (1999): 1-23.

*Mastrandrea, M. D., and S. H. Schneider, "Probabilistic Integrated Assessment of 'Dangerous' Climate Change." *Science* 304 (2004): 571-575

*McCarl, B. A., and R. D. Sands. "Competitiveness of Terrestrial Greenhouse Gas Offsets: Are They a Bridge to the Future?" *Climatic Change* 80 (2007): 109-126.

McFarland, J. R., J. M. Reilly, and H. J. Herzog. "Representing Energy Technologies in Top-down Economic Models Using Bottom-up Information." *Energy Economics*, vol. 26 (2004): 685-707.

*McKibbin, Warwick J., and Peter J. Wilcoxen. *Climate Change Policy after Kyoto: Blueprint for a Realistic Approach.* Washington, D.C.: Brookings Institution Press, 2002.

*McKibbin, Warwick J., and Peter J. Wilcoxen. "A Credible Foundation for Long Term International Cooperation on Climate Change." In *Architectures for Agreement: Addressing Global Climate Change in the Post-Kyoto World*, edited by Joseph Aldy and Robert Stavins. Cambridge University Press, 2007.

*McKibbin, Warwick J., and Peter J. Wilcoxen. "The Role of Economics in Climate Change Policy." *Journal of Economic Perspectives*, vol. 16, no. 2 (2002): 107-129.

Mendelsohn, Robert. "Efficient Adaptation to Climate Change." *Climatic Change*, vol. 45 (2000): 583-600.

Mendelsohn, Robert. "The Role of Markets and Governments in Helping Society Adapt to a Changing Climate." *Climatic Change*, vol. 78 (2006): 203-215.

Montgomery, W. David, and Anne E. Smith. *Price, Quantity, and Technology Strategies for Climate Change Policy.* Washington, D.C.: CRA International, October 11, 2005.

Morgan, Granger M., and Hadi Dowlatabadi. "Learning from Integrated Assessment of Climate Change." *Climatic Change*, vol. 34 (1996): 337-368.

Neumann, James, Gary Yohe, Robert Nicholls, and Michelle Manion. *Sea-Level Rise and Global Climate Change: A Review of Impacts to U.S. Coasts.* Arlington, Va.: Pew Center on Global Climate Change, February, 2000.

Newell, Richard G., Adam B. Jaffe, and Robert N. Stavins. "The Effects of Economic and Policy Incentives on Carbon Mitigation Technologies." *Energy Economics*, vol. 28 (2006): 563-578.

Newell, Richard, and William Pizer. "Discounting the Distant Future: How Much Do Uncertain Rates Increase Valuations?" Discussion paper 00-45. Washington, D.C.: Resources for the Future, 2000.

*Newell, R., Pizer, W., and Zhang, J. "Managing Permit Markets to Stabilize Prices." *Environmental and Resource Economics* 31 (2005): 133-157.

*Nordhaus, William D. "The Challenge of Global Warming: Economic Models and Environmental Policy." New Haven, Conn.: Yale University, July 24, 2007.

*Nordhaus, William D. "Critical Assumptions in the Stern Review on Climate Change." *Science*, vol. 317 (July 13, 2007): 201-202.

*Nordhaus, William D. "Economic Analyses of Kyoto Protocol: Is There Life After Kyoto?" Conference paper presented at Global Warming: Looking Beyond Kyoto, Yale University, New Haven, Conn., October 21-22, 2005.

*Nordhaus, William D. "Expert Opinion on Climatic Change." *American Scientist*, vol. 82. (1994): 45-51.

*Nordhaus, William D. "Geography and Macroeconomics: New Data and New Findings." *Proceedings of the National Academy of Sciences*, 103 (2006): 3510-3517.

Nordhaus, William D. *Managing the Global Commons: The Economics of Climate Change.* Cambridge: The MIT Press, 1994.

Nordhaus, William D. "An Optimal Transition Path for Controlling Greenhouse Gases." *Science*, vol. 258 (1992): 1315-1319.

Nordhaus, William D. *The "Stern Review" on the Economics of Climate Change*. Working Paper 12741. Cambridge: National Bureau of Economic Research, December 2006.

Nordhaus, William D. "To Slow or Not to Slow: The Economics of the Greenhouse Effect." *The Economic Journal*, vol. 101 (1991): 920-937.

*Nordhaus William D. "To Tax or Not to Tax? Alternative Approaches to Slowing Global Warming." *Review of Environmental Economics and Policy* 1(1) (Winter 2007): 26-44.

*Nordhaus, William D., and Joseph Boyer. *Warming the World: Economic Models of Global Warming*. Cambridge: the MIT Press, 2000.

Organisation for Economic Co-operation and Development. *The Benefits of Climate Change Policies*. Paris, France: 2004.

*Pacala, S., and R. Socolow. "Stabilization Wedges: Solving the Climate Problem for the Next 50 Years with Current Technologies." *Science* 305 (2004): 968-972.

*Paltsev, S., J. Reilly, H. Jacoby, A. Gurgel, G. Metċalf, A. Sokolov, and J. Holak. "Assessment of US Cap-and-Trade Proposals." MIT Joint Program on the Science and Policy of Global Change, Report 146 (2007).

UK Parliament House of Lords Select Committee on Economic Affairs. *The Economics of Climate Change*. Volume I: Report. HL-12-I. London, UK: 2005.

UK Parliament House of Lords Select Committee on Economic Affairs. *The Economics of Climate Change*. Vol. 2: Evidence. HL-12-II. London, UK: 2005.

UK Parliament House of Lords Select Committee on Economic Affairs. *Government Response to the Economics of Climate Change*. HL-71. London, UK: 2005.

Pearce, David. "The Political Economy of an Energy Tax: The United Kingdom's Climate Change Levy." *Energy Economics*, vol. 28 (2006): 149- 158.

Pearce, David. "The Role of Carbon Taxes in Adjusting to Global Warming." *The Economic Journal*, vol. 101, no. 407 (1991): 938-948.

Pearce, David. "The Social Cost of Carbon and its Policy Implications." *Oxford Review of Economic Policy*, vol. 19, no. 3 (2003): 362-384.

Peck, Stephen C., and Thomas J. Teisberg. "Optimal Carbon Emissions Trajectories when Damages Depend on the Rate or Level of Global Warming." *Climatic Change*, vol. 28 (1994): 289-314.

Pindyck, Robert S. *Uncertainty in Environmental Economics*. Washington, D.C.: AEI-Brookings Joint Center for Regulatory Studies, December 2006.

*Pizer, William A. "Combining Price and Quantity Controls to Mitigate Global Climate Change." *Journal of Public Economics* 85 (2002): 409-434.

*Pizer, William A. "The Evolution of a Global Climate Change Agreement." *American Economic Review* 96(2) (2006): 26-30.

Pizer, William A. "The Optimal Choice of Climate Change Policy in the Presence of Uncertainty." *Resource and Energy Economics*, vol. 21 (1999): 255-287.

Pizer, William A., Dallas Burtraw, Winston Harrington, Richard Newell, and James Sanchirico. *Modeling Economywide versus Sectoral Climate Policies Using Combined Aggregate-Sectoral Models*. Discussion Paper 05-08. Washington, D.C.: Resources for the Future, April, 2003.

*Popp, David. "R&D Subsidies and Climate Policy: Is There a 'Free Lunch'?" *Climatic Change* 77(3-4) (2006): 311-341.

Portney, Paul R., and John P. Weyant, Editors. *Discounting and Intergenerational Equity*. Washington D.C.: Resources for the Future, 1999.

Reilly, John M., Editor. *Agriculture: The Potential Consequences of Climate Variability and Change for the United States*. Cambridge, UK: Cambridge University Press, 2002.

*Reilly, J., and M. Asadoorian. "Mitigation of Greenhouse Gas Emissions from Land Use: Creating Incentives Within Greenhouse Gas Emissions Trading Systems." *Climatic Change* 80 (2007): 173-197.

*Reilly J., S. Paltsev, B. Felzer, X. Wang, D. Kicklighter, J. Melillo, R. Prinn, M. Sarofim, A. Sokolov, and C. Wang. "Global Economic Effects of
Changes in Crops, Pasture, and Forests Due to Changing Climate, Carbon Dioxide, and Ozone." MIT Joint Program on the Science and Policy of Global Change, Report No. 149, May 2007.

*Reilly, J., M. Sarofim, S. Paltsev, and R. Prinn. "The Role of Non-CO2 Greenhouse Gases in Climate Policy: Analysis Using the MIT IGSM." *Energy Journal*, Special Issue on Multigas Mitigation and Climate Policy (2006): 503-520.

Riahi, Keywan, Edward S. Rubin, Margaret R. Taylor, Leo Schrattenholzer, and David Hounshell. "Technological Learning for Carbon Capture and Sequestration Technologies." *Energy Economics*, vol. 26, (2004): 539-564.

Richels, Richard, and Jae Edmonds. "The Economics of Stabilizing Atmospheric CO2 Concentrations." *Energy Policy*, vol. 23, no. 4/5 (1995): 373-378.

Richels, Richard, Thomas Rutherford, Geoffrey Blanford, and Leon Clarke. *Managing the Transition to Climate Stabilization*. Working Paper 07-01.

Washington, D.C.: AEI-Brookings Joint Center for Regulatory Affairs, January 2007.

Schelling, Thomas C. "Intergenerational Discounting." *Energy Policy*, vol. 23, no. 4/5 (1995): 395-401.

Schneider, Stephen, and Janica Lane. "Integrated Assessment Modeling of Global Climate Change: Much Has Been Learned — Still a Long and Bumpy Road Ahead." *The Integrated Assessment Journal*, vol. 5, iss. 1 (2005): 41- 75.

Smith, Joel B. *A Synthesis of Potential Climate Change Impacts* on the U.S. Arlington, Va.: Pew Center on Global Climate Change, April, 2004.

Sohngen, Brent, Roger Sedjo, Robert Mendelsohn, and Ken Lyon. *Analyzing the Economic Impact of Climate Change on Global Timber Markets*. Discussion Paper 96-08. Washington, D.C.: Resources for the Future, 1996.

*Stavins, Robert N. "Proposal for a U.S. Cap-and-Trade System to Address Global Climate Change: A Sensible and Practical Approach to Reduce Greenhouse Gas Emissions." The Hamilton Project, the Brookings Institution, Washington, D.C.: September 12, 2007.

Stavins, Robert N., Judson Jaffe, and Todd Schatzki. *Too Good to be True? An Examination of Three Economic Assessments of California Climate Change Policy*. Related Publication 07-01. Washington, D.C.: AEI- Brookings Joint Center for Regulatory Affairs, January, 2007.

Stavins, Robert N., and Kenneth R. Richards. *The Cost of U.S. Forest- Based Carbon Sequestration*. Arlington, Va.: Pew Center on Global Climate Change, 2005.

Stern, Nicholas. *The Economics of Climate Change*: The Stern Review. Cambridge, UK: Cambridge Press, 2007.

Sutherland, Ronald J. "'No Cost' Efforts to Reduce Carbon Emissions in the U.S.: An Economic Perspective." *Energy Journal*, vol. 21, no. 3 (2000): 89-1 12.

*Tavoni, M., Brent Sohngen, and Valentina Bosetti. "Forestry and Carbon Market Response to Stabilize Climate." *Energy Policy*, 35 (2007), 5346– 5353.

*Thomas, C. D., et al. "Extinction Risk from Climate Change." *Nature* 247 (2004): 145-148.

Tol, Richard S. J. "The Damage Costs of Climate Change Toward More Comprehensive Calculations." *Environmental and Resource Economics*, vol. 5 (1995): 353-374.

Tol, Richard S. J. "Europe's Long-Term Climate Target: A Critical Evaluation." *Energy Policy*, vol. 35, no. 1 (2007): 424-432.

Tol, Richard S. J. "Is the Uncertainty About Climate Change Too Large for Expected Cost-Benefit Analysis?" *Climatic Change*, vol. 56 (2003): 265- 289.

Tol, Richard S. J. "The Marginal Costs of Greenhouse Gas Emissions." *The Energy Journal*, vol. 20, no. 1 (1999): 61-81.

*Tol, Richard S. J. "The Marginal Damage Costs of Carbon Dioxide Emissions: An Assessment of the Uncertainties." *Energy Policy*, 33 (2005): 2064-2074.

Tol, Richard S. J. "Welfare Specifications and Optimal Control of Climate Change: An Application of FUND." *Energy Economics*, vol. 24 (2002): 367-376.

Tol, Richard S. J., and Gary Yohe. "A Review of the Stern Review." *World Economics*, vol. 7, no. 4 (2006): 233-250.

Van Vuuren, Detlef, John Weyant, and Francisco de la Chesnaye. "Multi- Gas Scenarios to Stabilize Radiative Forcing." *Energy Economics*, vol. 28 (2006): 102-120.

Webster, Mort D., Mustafa H. Babiker, Monika Mayer, John M. Reilly, Jochen Harnisch, Robert Hyman, Marcus C. Sarofim, and Chien Wang. *Uncertainty in Emissions Projections for Climate Models*. Report No. 79. Cambridge: MIT Joint Program on the Science and Policy of Global Change, August, 2001.

*Webster, Mort, C. Forest, J. Reilly, M. Babiker, D. Kicklighter, M. Mayer, R. Prinn, M. Sarofim, A. Sokolov, P. Stone, and C. Wang. "Uncertainty Analysis of Climate Change and Policy Response." *Climatic Change*, 61 (2003): 295-320.

Weitzman, Martin L. "Prices versus Quantities." *The Review of Economic Studies*, vol. 41, no. 4 (1974): 477-491.

*Weitzman, Martin L. "The Role of Uncertainty in the Economics of Catastrophic Climate Change." AEI-Brookings Joint Center for Regulatory Studies, Working Paper 07-11, May 2007.

Weitzman, Martin L. *Structural Uncertainty and the Value of Statistical Life in the Economics of Catastrophic Climate Change*. Working Paper 07-11. Washington, D.C.: AEI-Brookings Joint Center for Regulatory Studies, Revised October 2007.

Weyant, John P. "Introduction and Overview." *Energy Economics*, vol. 26 (2004): 501-515.

Wigley, T.M.L. "The Kyoto Protocol: CO2, CH4 and Climate Implications." *Geophysical Research Letters* 25 (1998): 2285–2288.

Wigley, T. M. L., R. Richels, and J. A. Edmonds. "Economic and Environmental Choices in the Stabilization of Atmospheric CO2 Concentrations." *Nature*, vol. 379 (1996): 240-243.

Wing, Ian Sue. "Representing Induced Technological Change in Models for Climate Policy Analysis." *Energy Economics*, vol. 28 (2006): 539-562.

*Yohe, G. "Carbon Emissions Taxes: Their Comparative Advantage under Uncertainty." *Annual Review of Energy* 17 (1992): 301-326.

*Yohe, G. "More on the Properties of a Tax Cum Subsidy Pollution Control Strategy." *Economic Letters* 31 (1989): 193-198.

Yohe, Gary. "Some Thoughts on the Damage Estimates Presented in the Stern Review — An Editorial." *The Integrated Assessment Journal,* vol. 6, iss. 3 (2006): 65-72.

*Yohe, Gary, Natasha Andronova, and Michael Schlesinger. "To Hedge or Not Against an Uncertain Climate Future?" *Science,* vol. 306 (2004): 416- 417.

Yohe, Gary, and Michael E. Schlesinger. "Sea-level change: The Expected Economic Cost of Protection or Abandonment in the United States." *Climatic Change,* vol. 38 (1998): 447-472.

Yohe, G., M. E. Schlesinger, and N. G. Andronova. "Reducing the Risk of a Collapse of the Atlantic Thermohaline Circulation." *The Integrated Assessment Journal,* vol. 6 (2006): 57-73.

*Yohe, G. W., R. S. J. Tol, and D. Murphy. "On Setting Near-term Climate Policy while the Dust Begins to Settle: The Legacy of the Stern Review." *Energy and Environment* 18 (2007): 621-633.

ENDNOTES

[1] Major greenhouse gases include carbon dioxide (CO_2), methane (CH_4), nitrous oxide (N_2O), and synthetic gases (hydrofluorocarbons {HFCs}, perfluorocarbons {PFCs}, and sulfur hexafluoride {SF_6}).

[2] See GAO,*Climate Change: Financial Risks to Federal and Private Insurers in Coming Decades Are Potentially Significant*, GAO-07-285, Mar. 16, 2007; and *Climate Change: Agencies Should Develop Guidance for Addressing the Effects on Federal Land and Water Resources*, GAO-07-863, (Washington, D.C.: Aug. 7, 2007).

[3] Atmospheric concentrations of carbon dioxide increased from 280 parts per million to 379 parts per million between pre-industrial times and 2005. (See Intergovernmental Panel on Climate Change (IPCC), 2007: The Physical Science Basis. Contribution of Working Group I to the Fourth Assessment)

[4] Environmental Protection Agency (EPA), 2008, *Inventory of U.S. Greenhouse Gas Emissions and Sinks, 1990-2006,* Public Review Draft (Washington, D.C.: Apr. 18, 2008).

[5] Energy Information Administration (EIA), *International Energy Annual 2005*.

[6] In the context of the United Nation's Framework Convention on Climate Change, mitigation is a human intervention to reduce the sources or enhance the sinks of greenhouse gases. Examples include using fossil fuels more efficiently for industrial processes or electricity generation, switching to solar energy or wind power, improving the insulation of buildings, and expanding forests and other sinks to remove greater amounts of carbon dioxide from the atmosphere. Forests and other vegetation are considered sinks because they remove carbon dioxide through photosynthesis.

[7] In general, auctioning the allowances would enable the government to decide how to use the revenue, and allocating the allowances for free would represent a transfer of wealth from the government to the entities receiving the allowances.

[8] In general, a firm will purchase permits when the permit price is lower than the cost to abate emissions.

[9] Establishing a price on emissions creates a price signal to fossil fuel users to cut back on consumption.

[10] National Oceanic and Atmospheric Administration (NOAA), "Coastal Area and Marine Resources: The Potential Consequences of Climate Variability and Change," December 2001.

[11] The appropriate price, often called the social cost of carbon, reflects the present value of economic damages caused by an additional quantity of greenhouse gas emissions. Under an economically optimal control policy, the price would be set at the point where the incremental or marginal damages from global warming equal the marginal costs of controlling emissions.

[12] International Panel on Climate Change, Working Group II (2007). A metric ton is equivalent to 1,000 kilograms, or approximately 2,204 pounds.

[13] One panelist preferred a policy that set a price for each individual greenhouse gas as opposed to one price for all emissions.

[14] Because greenhouse gases differ in their potential to contribute to global warming, each gas is assigned a unique weight, called a global warming potential, based on its heat-absorbing ability relative to carbon dioxide over a fixed period. This provides a way to convert emissions of various greenhouse gases into a common measure, called carbon dioxide equivalent.

[15] IPCC, 2007: Introduction. In: *Climate Change 2007: Mitigation, Contribution of Working Group III to the Fourth Assessment Report of the Intergovernmental Panel on Climate Change* (B. Metz, O.R. Davidson, P.R. Bosch, R. Dave, L.A. Meyer [eds.]), Cambridge University Press, Cambridge, United Kingdom.

[16] IPCC, 2007: Introduction. In: *Climate Change 2007: Mitigation, Contribution of Working Group III to the Fourth Assessment Report of the Intergovernmental Panel on Climate Change* (B. Metz, O.R. Davidson, P.R. Bosch, R. Dave, L.A. Meyer [eds.]), Cambridge University Press, Cambridge, United Kingdom.

[17] The IPCC defines mitigation as technological change and substitution that reduce resource inputs, such as energy use, and emissions per unit of output. Although several social, economic, and technological policies would produce an emissions reduction, with respect to climate change, mitigation means implementing policies to reduce greenhouse gas emissions and enhance greenhouse gas sinks.

[18] UNFCCC, Bali Action Plan, December 2007.

[19] National Research Council: *Abrupt Climate Change: Inevitable Surprises* (Washington, D.C.: 2002).

[20] In general, benefits and costs are measured in terms of a common metric: dollars. Benefits are monetized by estimating the amount that individuals would be willing to pay for the benefit (or the amount they would be willing to accept to forgo the benefit) based on market transactions. In cases where the benefit is not traded in a market (e.g., ecosystem services), economists use methods such as contingent valuation surveys to elicit the amount that individuals would be willing to pay for the benefit.

[21] A recent report sponsored by the British government (i.e., the Stern Review) concluded that relatively aggressive emissions reductions by the global community were economically justified. Several economists criticized its discounting approach. See appendix V for literature citations on this topic.

[22] IPCC, 2007: Summary for Policymakers. In *Climate Change 2007: Impacts, Adaptation and Vulnerability. Contribution of Working Group II to the Fourth Assessment Report of the Intergovernmental Panel on Climate Change*, M. L. Parry, O. F. Conziani, J. P. Palutikof, P. J. van der Linden, and C. E. Hanson, Eds., Cambridge University Press, Cambridge, United Kingdom, 7-22.

[23] S. 280 (Lieberman), S. 309 (Sanders), S. 317 (Feinstein), S. 485 (Kerry), S. 1766 (Bingaman), S. 2191 (Lieberman/Warner), H.R. 620 (Olver), and H.R. 1590 (Waxman).

[24] The remaining 4 panelists did not provide a response.

[25] In specifying an initial price for a cap-and-trade or hybrid program, the emissions cap could be set at a level that would be consistent with the initial price (i.e., the targeted permit price).

[26] One panelist preferred a policy that set a price for each individual greenhouse gas as opposed to one price for all emissions.

[27] Several noted that some categories of benefits, such as establishing a price signal, are not benefits per se because they are not the result of reducing emissions. Nonetheless, because these categories were identified by other panelists as benefits of actions to address climate change, we asked all the panelists to rate their importance.

[28] CCSP cited (1) the amount that emissions must be reduced to achieve an emissions path to stabilization, and (2) the technologies that are available to facilitate changes in the economy as reasons for the difference in stabilization costs between models.

[29] The science of climate change refers to knowledge about the effect on the climate of factors such as greenhouse gas concentrations, clouds, and the carbon cycle.

[30] The science of climate change refers to knowledge about the effect on the climate of factors such as greenhouse gas concentrations, clouds, and the carbon cycle.

INDEX

A

abatement, 28, 31
academic, xv, 23, 39
access, 41
acid, xiii, 32
adaptability, 31
adaptation, 1, 3, 9, 13, 20, 26, 34, 35, 36, 43, 44, 45, 52, 54
administration, 14
administrative, 20
AEI, 66, 67, 68
Africa, 9, 16
agricultural, xi, xv, 9, 10, 15, 44, 48
agriculture, xii, 9, 20
aid, 20
air, xii, 5, 7
air quality, 7
alternative, 2, 25, 31, 36, 45, 48, 52
alternative energy, 45
Amsterdam, 61
anthropogenic, xii, 47
appendix, 20, 21, 22, 24, 27, 30, 33, 41, 73
Arctic, 9
Asia, 10
assessment, 2, 14, 17, 23, 25, 26, 40, 41, 53, 54
assessment models, 2, 14, 17, 25, 26, 41, 53, 54
assumptions, xiv, xv, 25, 41, 53

Atlantic, 69
atmosphere, vii, xi, 5, 11, 15, 20, 71
Australia, 10
availability, 6, 9, 10, 41

B

Bali, xii, 13, 72
barriers, 9, 11
behavior, 23, 53
benefits, xiv, xv, 1, 2, 14, 15, 17, 20, 22, 23, 25, 26, 29, 30, 31, 32, 35, 36, 40, 41, 49, 50, 51, 52, 53, 54, 73
bias, xv
binding, xii
biodiversity, 9, 10
Brazil, 28
buildings, 35, 71
burning, xii

C

campaigns, 11
capacity, 9, 13, 16, 36
capacity building, 9, 13
caps, 16, 31
carbon, xi, xii, xiii, xv, 5, 7, 11, 16, 18, 19, 20, 21, 23, 24, 32, 34, 43, 44, 45, 46, 48, 51, 53, 59, 60, 61, 63, 64, 65, 66, 67, 68, 69, 71, 72, 74
CEA, 39, 42

CH4, 68, 71
chemicals, 28
China, xii, 28, 44
climate change, vii, xi, xii, xiii, xiv, xv, 2, 7, 9, 11, 13, 14, 15, 16, 17, 18, 19, 20, 21, 22, 23, 25, 26, 29, 31, 32, 34, 35, 36, 39, 40, 41, 47, 48, 49, 52, 53, 55, 72, 73, 74
Climate Change Science Program (CCSP), ix, xii, 23, 60, 73
clouds, 74
Co, 63
CO2, 61, 63, 66, 68, 71
coal, vii, xii, 28
coal mine, xii
coastal areas, xi, xiv, 9, 10
coastal communities, 13
combustion, vii, xiv
commerce, 7
Committee on Environment and Public Works, xi
communities, 9, 13, 23
community, 73
compatibility, 32
compensation, 21
competition, 9
competitiveness, 28
complement, 17, 29, 33, 35
complex interactions, xi
compliance, 11, 14
components, 2, 19, 32
concentration, 14, 24
conditioning, xii
confidence, 15
conflict, 53
conformity, 40
Congress, xiii, xv, 1, 2, 16, 17, 18, 20, 26, 29, 31, 34, 36, 41, 47, 49, 51, 54
Congressional Budget Office, 60
Connecticut, 57, 59
constraints, 13
consumers, xiii, 3, 11, 15, 32, 53
consumption, 9, 15, 27, 55, 72
content analysis, 41
control, 1, 3, 14, 18, 19, 48, 72
cooling, 7

coral, 6, 9
coral reefs, 6, 9
cost effectiveness, 31
cost-effective, 2, 14, 34, 48
costs, xiv, xv, 1, 2, 7, 9, 11, 14, 15, 16, 17, 19, 20, 21, 22, 23, 25, 26, 29, 30, 31, 32, 34, 35, 36, 40, 41, 49, 50, 51, 53, 54, 72, 73
Council on Environmental Quality, 42
coverage, 7
covering, 11
credibility, 2, 28, 47
crops, 6

D

Dallas, 66
danger, 6
Dartmouth College, 57
deaths, 6, 7
decision makers, 2, 26, 32, 49
decision making, 17, 25, 39
degradation, 6
Delphi, xv, 39, 41
demand, xiii, 6, 7, 10
Department of Agriculture, ix, xv, 39, 63
Department of Commerce, xv, 39
Department of Energy (DOE), ix, xiii, xv, 39
developed countries, 7, 11
developing countries, 7, 10, 20, 45, 48, 53
discount rate, 15, 25, 53
discounting, 15, 25, 73
diseases, 15
distribution, 3, 16, 31, 51
domestic policy, 33
dominance, 40
drought, 10
droughts, 10
drowning, 7

E

earth, 5
earth's atmosphere, xi
Eastern Europe, 10

Index

ecological, xiv
ecological economics, 59
economic damages, 72
economic development, xii
economic efficiency, 14, 29, 31
economic growth, xiv, 5, 23, 54
Economic Research Service, 63
economic theory, xiv
economically disadvantaged, 36
economics, xv, 26, 41
ecosystem, 73
ecosystems, xii, xiv, 5, 6, 9, 10, 15, 22, 23, 52
electric utilities, xiii, 20
electricity, 16, 20, 24, 28, 71
emission, 11, 52
emitters, xiii, 19, 28, 32, 44
encapsulated, 41
energy, xi, xiii, xiv, 3, 7, 9, 11, 12, 13, 14, 16, 24, 26, 28, 34, 35, 36, 43, 44, 45, 48, 54, 71, 72
energy efficiency, xiii, xiv, 3, 11, 16, 43, 44
Energy Information Administration (EIA), xv, 25, 39, 60, 71
energy supply, xiv
engagement, 45
environment, 28
Environmental Protection Agency, ix, xiii, xv, 39, 42, 60, 62, 71
equilibrium, 24
equity, 31, 36
erosion, 9, 10
estimating, 1, 15, 17, 25, 26, 53, 73
estuaries, 6
Europe, 10
European Union, ix, xii, xiii, 11, 32
excuse, 36
expenditures, 14
expert, 24, 29, 39
expertise, xv, 40
exposure, 6, 9

F

federal government, xi, 34, 36, 40

feedback, xv, 39
fertilizers, xii, 12
fire, 9
fires, 10
firms, xiii, 18, 20
fish, 9
fisheries, 9, 20
fishing, 7
flexibility, 18, 31, 32, 39
flood, xi, 7, 48
flooding, 2, 7, 9, 10, 22, 23, 36, 52
fluctuations, 2
food, xii, 6, 9, 10, 16
food production, xii
forestry, 6, 10, 20, 67
forests, 71
fossil fuels, vii, xii, xiii, xiv, 19, 71
France, 62, 65
freshwater, 6, 9, 10
fuel, xiii, 3, 12, 19, 34, 48, 72
fuel efficiency, 34
funding, 20, 26, 34, 36, 44

G

gas, vii, xii, xiii, xiv, 11, 14, 15, 16, 17, 18, 28, 72
gases, xi, xii, 5, 14, 15, 43, 44, 45, 46, 71, 72
gasoline, 24
generation, 7, 9, 11, 20, 71
GHG, 63
glaciers, 5, 9
global warming, 15, 72
Global Warming, 60, 64, 65
goals, 14, 18, 32, 35, 40
goods and services, xiv, 14, 15
government, xi, xii, xiii, xv, 1, 2, 19, 28, 32, 34, 35, 36, 40, 71, 73
greenhouse, vii, xi, xii, xiii, xiv, 1, 2, 5, 7, 11, 13, 14, 15, 16, 17, 18, 19, 20, 23, 25, 27, 28, 31, 32, 43, 44, 46, 48, 51, 52, 53, 54, 59, 60, 61, 62, 63, 65, 66, 67, 68, 71, 72, 73, 74
greenhouse gas, vii, xi, xii, xiii, xiv, 1, 2, 5, 7, 11, 13, 14, 15, 16, 17, 18, 19, 20, 23, 25,

27, 28, 31, 32, 43, 44, 46, 48, 51, 52, 53, 54, 71, 72, 73, 74
greenhouse gases, vii, xi, xiii, 5, 14, 15, 18, 20, 23, 28, 31, 48, 51, 54, 71, 72
Greenland, 5
gross domestic product, xiv
groundwater, 6
groups, 45, 49, 50, 51
growth, xiv, 5, 9, 10, 23, 54

H

Harvard, 57
hazards, 9
health, xi, 6, 10
heat, 5, 6, 72
heating, 7, 9, 10
high temperature, 26
House, 63, 65
human, 5, 6, 9, 14, 15, 71
human welfare, 14, 15
humans, 6
hurricanes, 36
hybrid, 2, 18, 19, 73
hydropower, 7

I

ice, 5, 7, 9
implementation, 18, 43
incentive, xiii, 34
incentives, 11, 12, 13, 23, 34, 36, 53
incidence, 15
inclusion, 25
income, 27, 32, 49, 50, 51
India, xii, 28, 44
indigenous, 9
industrial, xii, 11, 23, 71
industry, xiv, 7, 16
infectious, 6
inflation, 18, 19, 25
information sharing, xiii
infrastructure, vii, xi, xiv, 7, 9, 12, 13, 36
injuries, 6, 7

innovation, 34, 62
insecurity, 16
institutions, xv, 40
instruments, 11, 12
insulation, 71
insurance, xi, 36
integration, 13
intensity, vii, xi, 9
interactions, xi
Intergovernmental Panel on Climate Change, (IPCC) ix, xi, xiv, 5, 7, 8, 13, 16, 61, 62, 71, 72, 73
international trade, 28
internet, 39, 41
intervention, 34, 71
investment, xiv, 12, 13, 20, 28, 35, 53
iron, xii
irrigation, 6
island, 7, 9

J

Japan, xii
judgment, 26

K

Kyoto Protocol, xii, 11, 62, 63, 64, 68

L

labeling, 12
labor, 51
land, 7, 12, 34, 45, 46, 48
land use, 7, 45, 46, 48
landfills, xii
Latin America, 9
lead, xii, 2, 13
leadership, 12
leakage, 48, 50
light trucks, 48
likelihood, 35, 53
limitation, 17, 40

Index

limitations, xv, 2, 3, 29, 34, 36, 40
livestock, 6
local government, 34
location, 9, 11
London, 60, 65
long period, 15
long-term, xi, 18, 34, 35
loopholes, 45
losses, 5
low-income, 32, 51
lying, 9, 10

M

malnutrition, 6
management, 12, 13, 34, 36
manufacturing, xii, 11
marginal costs, 31, 32, 72
marginal utility, 27, 55
market, xiii, xiv, 1, 2, 17, 18, 19, 23, 25, 29, 31, 32, 33, 35, 43, 53, 73
market failure, 35
markets, 11, 15
Massachusetts Institute of Technology (MIT), 61, 62, 64, 65, 66, 68
measures, xiv, 12
Medicare, 52
melting, 5
metals, 28
methane, xi, xii, 5, 71
metric, xv, 18, 19, 46, 72, 73
migration, 7
missions, ix, 7, 11, 32, 48
mitigation policy, xiii, 34, 35, 36
modeling, 23, 40
models, xiv, 2, 14, 17, 25, 26, 41, 53, 54, 73
moderates, 5
mortality, 6
mountains, 9
movement, 7

N

NAS, ix, xv, 15, 40
NASA, ix, xv, 39
nation, 43, 44, 45
national, 11
National Academy of Sciences (NAS), ix, xv, 40, 64
National Aeronautics and Space Administration, ix, xv
National Oceanic and Atmospheric Administration (NOAA), ix, xv, 39, 72
National Research Council, 73
native species, 9
natural, vii, xi, xv, 9, 13, 36, 53
natural disasters, 36
natural gas, vii
natural resources, xi, 53
negotiation, 44
New Zealand, 10
nitrous oxide, xii, 5, 71
non-native, 9
non-native species, 9
North America, 9

O

oceans, xi, 5
Ohio, 57
oil, vii, xii, 11
oil refineries, 11
Organisation for Economic Co-operation and Development, 65
organization, xii, 35
oversight, 48
oxide, xii, 5, 71
ozone, 66

P

pacific, 9, 57
Paper, 59, 61, 63, 64, 65, 66, 67, 68
Paris, 62, 65
Parliament, 65
password, 41
peatland, 10
Pennsylvania, 39

performance, 16
periodic, 19
permit, 3, 19, 31, 72, 73
photosynthesis, 71
planning, vii, 12, 13
planning decisions, 13
plants, 11
polar ice caps, 5
policy choice, 23
politics, 34
pollutant, 32
pollutants, xiii
poor, 7, 13
population, xiv, 7, 9, 10, 54
population growth, 9, 10, 54
portfolio, 1, 11, 16, 17, 20, 43, 47
poverty, 13, 16
power, 11, 71
precipitation, 10
preference, 27, 55
present value, 14, 15, 72
pressure, 40
prices, xiii, xiv, 14, 24, 26, 32
private, xiii, 7, 20, 44, 48
private sector, 20, 44, 48
probability, 23, 26
probability distribution, 26
producers, xiii, xiv, 11, 15, 19
production, xii, 3, 10, 28, 30
productivity, xi, 9, 10, 15
program, xii, xiii, 2, 17, 18, 19, 23, 24, 28, 31, 32, 36, 51, 73
promote, 16
property, iv, 36
property rights, 36
protection, 7, 36
public, xiii, xiv, 6, 12, 20, 22, 23, 35, 48, 52, 54
public health, xiv, 22, 23, 48, 52, 54
public investment, 20, 35
public sector, xiii

Q

questionnaire, 20, 22, 23, 25, 26, 31, 33, 39, 40, 41
questionnaires, xv, 39, 40, 41

R

R&D, 34, 35, 43, 44, 45, 46, 48, 51, 52, 60, 62, 66
radiation, 59
rain, xiii, 9, 32
range, xv, 1, 18, 21, 22, 24, 27, 30, 33, 40, 43, 46
ratings, 20
recovery, 11
reduction, 11, 13, 14, 18, 23, 24, 28, 45, 52, 72
reefs, 6, 9
reflectivity, xi
refrigeration, xii
regional, 15
regulation, vii, xiii, 43, 46
renewable energy, xiv, 12, 16
rent, 35, 50
research, xiii, xiv, xv, 1, 3, 11, 12, 13, 19, 20, 23, 26, 32, 34, 35, 36, 40, 44, 45
research and development, xiv, 1, 3, 11, 20, 32, 34, 35, 36, 44, 45
researchers, 14, 15, 26
residential, 24
resilience, 16
resource management, 36
resources, xi, 6, 7, 9, 13, 14, 34, 35, 36, 53
respiratory, 6
revenue, xiii, 3, 32, 51, 54, 71
risk, 6, 7, 9, 10, 13, 16, 23, 26, 36, 53, 54
risk management, 13, 36
risks, 2, 10, 23, 26
rural, 7
Russia, xii
Rutherford, 66

Index

S

safety, 1, 17, 18, 19, 31, 32, 44, 45
salinization, 6
saltwater, 6
scarcity, 6, 9, 10, 16, 23
scientific understanding, xii, 52
scientists, xiv, 15
sea level, vii, xi, xiv, 2, 5, 7, 13, 23
sea-level, 10
sea-level rise, 10
seawater, 5
Secretary of Commerce, xii
security, 9, 16
semiconductor, xii
Senate, xi
sensitivity, xiv, 24, 54
series, 39, 40
services, iv, xiv, 14, 15, 73
settlements, 7, 9
severity, 10
sharing, xiii
shortage, 6
sites, 10
skin, 6
skin diseases, 6
social costs, xiv
Social Security, 52
social welfare, 23
soils, 6
solar, 16, 71
solar energy, 16, 71
solutions, 32
South America, 11
South Asia, 10
Southeast Asia, 10
species, 9, 10, 22, 23, 52
sponsor, 45
stabilization, 14, 23, 73
stabilize, xii, 14, 15
standards, xiii, 3, 11, 12, 16, 34, 35, 43, 48
steel, xii, 11
stochastic, 59
storage, 11, 20, 32, 44, 45, 46, 51
storms, vii, 10

strategies, 13, 26, 45
subsidies, 3, 11, 12, 34, 35, 36, 43, 44, 48
subsidy, 35, 36, 48
substances, 5
substitution, 72
sulfur, 32, 71
sulfur dioxide, 32
summer, 9, 10
supply, xiii, 6
susceptibility, 13
switching, 71
systems, 6

T

targets, 2, 19, 20, 32, 35
tariffs, 44
tax credit, 11, 12
tax credits, 11, 12
taxes, 2, 11, 19, 51, 54
technical assistance, xiii
technological change, xiv, 54, 72
technology, 11, 34, 45, 48, 50
technology transfer, 34, 45
temperature, 5, 14, 15, 24, 26
territory, 53
Texas, 57
threatened, xii
threatening, 9
thresholds, 26, 50
time frame, xii
time periods, 15
timing, xiv
total costs, 14
tourism, 7, 9, 10
trade, xiii, xv, 1, 2, 11, 14, 15, 16, 17, 18, 19, 23, 28, 29, 31, 32, 33, 40, 43, 44, 45, 47, 48, 73
trade-off, xv, 29, 40
trading, xiii, 32
transactions, 73
transfer, 34, 45, 71
transport, 7
transportation, 16
trees, 6

U

tropical forest, 9
tropical storms, xi
trucks, 48

U.S. Department of Agriculture, 63
UK, 60, 61, 62, 65, 66, 67
uncertainty, 2, 15, 26, 32, 34, 35, 36, 50, 52, 54
UNFCCC, ix, 72
United Kingdom, 65, 72, 73
United Nations, ix, xii
United States, vi, vii, xi, xii, xiii, xv, 2, 7, 11, 13, 17, 18, 20, 23, 27, 28, 32, 39, 49, 50, 59, 60, 66, 69
urbanization, 10
USDA, ix, xv, 39

V

values, 14, 47
variability, 9

vegetation, 71
vehicles, 34, 48
vineyards, 59
visible, 30
volatility, 31, 32
vulnerability, 7, 9, 10, 13, 16, 23

W

wastewater, 13
water, 6, 9, 10, 16, 23, 34, 36
water resources, 6, 9, 34
wealth, 71
welfare, 23, 27
wildfires, 6
wind, 16, 71
winning, 36
winter, 9, 10

Y

yield, 10, 34